John Croumbie Brown

Forests and Forestry of Northern Russia and Lands Beyond

John Croumbie Brown

Forests and Forestry of Northern Russia and Lands Beyond

ISBN/EAN: 9783337336974

Printed in Europe, USA, Canada, Australia, Japan

Cover: Foto ©berggeist007 / pixelio.de

More available books at **www.hansebooks.com**

FORESTS AND FORESTRY

OF

NORTHERN RUSSIA

AND LANDS BEYOND.

COMPILED BY

JOHN CROUMBIE BROWN, LL.D.,

Formerly Lecturer on Botany in University and King's College, Aberdeen; subsequently Colonial Botanist at Cape of Good Hope, and Professor of Botany in the South African College, Capetown; Fellow of the Linnean Society; Fellow of the Royal Geographical Society; and Honorary Vice-President of the African Institute of Paris.

EDINBURGH:
OLIVER AND BOYD, TWEEDDALE COURT.
LONDON: SIMPKIN, MARSHALL, & CO.,
AND WILLIAM RIDER & SON.
MONTREAL: DAWSON BROTHERS.

1884.

ADVERTISEMENT.

In the Spring of 1877 I published a *Brochure* entitled *The Schools of Forestry in Europe: a Plea for the Creation of a School of Forestry in connection with the Arboretum in Edinburgh*, in which, with details of the arrangements made for instruction in Forest Science in Schools of Forestry in Prussia, Saxony, Hanover, Hesse, Darmstadt, Wurtemburg, Bavaria, Austria, Poland, Russia, Finland, Sweden, France, Italy, and Spain, and details of arrangements existing in Edinburgh for instruction in most of the subjects included amongst preliminary studies, I submitted for consideration the opinion, 'that with the acquisition of this Arboretum, and with the existing arrangements for study in the University of Edinburgh, and in the Watt Institution and School of Arts, there are required only facilities for the study of what is known on the Continent as Forest Science to enable these Institutions conjointly, or any one of them, with the help of the others, to take a place amongst the most completely equipped Schools of Forestry in Europe, and to undertake the training of foresters for the discharge of such duties as are now required of them in India, in our Colonies, and at home.'

This month has seen an International Exhibition of forest products, and other objects of interest connected with forestry, opened in Edinburgh, 'in the interests of forestry, and to promote a movement for the establishment of a School of Forestry in Scotland, as well as with a view of furthering and stimulating a greater improvement in the scientific management of woods in Scotland and the sister countries which has manifested itself during recent years.'

The following is one of a series of volumes published in support of this enterprise, with a view to introduce into English forestal literature detailed information on some of the points on which information is supplied to the students at Schools of Forestry on the Continent; and to make better known the breadth of study which is embraced in what is known there as *Forstwissenschaft*, or Forest Science.

JOHN C. BROWN.

HADDINGTON, *24th July, 1884.*

CONTENTS.

PART I.—FOREST LANDS.

	PAGE
INTRODUCTION,	1
CHAPTER I.—*The Neva*,	3
Voyage from St. Petersburg to Schlusselburg.	
CHAPTER II.—*Lake Ladoga*,	11
CHAPTER III.—*The Svir*,	16
CHAPTER IV.—*Lake Onega*,	22
CHAPTER V.—*The Falls of Keewash*,	27
CHAPTER VI.—*Forest Lands of Olonetz*,	36
CHAPTER VII.—*Forests of Archangel*,	49
CHAPTER VIII.—*Lapland, and Land of the Samoides*,	59
CHAPTER IX.—*Nova Zembla, and Lands beyond*,	73

PART II.—FOREST EXPLOITATION.

CHAPTER I.—*Sartage*,	85

	PAGE
CHAPTER II.—*Jardinage,*	89

Disastrous Consequences at Cape of Good Hope (p. 90), and Elsewhere (p. 95); Precautions adopted here (p. 96).

CHAPTER III.—*Views entertained in Russia in regard to different Methods of Exploitation,*	101

CHAPTER IV.—*Export Timber Trade,*	109

Onega or English Timber Company (p. 109); Transport and Floatage (p. 112); Cutting up of Logs (p. 115); Other Companies (p. 117); Forest Code (p. 123).

CHAPTER V.—*Exports by Archangel and the White Sea,*	125
CHAPTER VI.—*Forest Industries,*	133
SECTION A. - *Forest Exploitation and Clearing of Forest Lands,*	133
SECTION B.—*Tar, Turpentine, and Vinegar Manufacture,*	137
SECTION C.—*House Building and Carpentry,*	141

———o———

PART III.—PHYSICAL GEOGRAPHY.

CHAPTER I.—*Contour and General Appearance of the Country,*	143

Rivers (p. 144); Lapland and its Temperature (p. 147).

CHAPTER II.—*Flora,*	155

CONTENTS.

                                                                                PAGE

SECTION 1.—*Characteristic Vegetation,* - - 155
Successive Zones of Vegetation characteristic of Latitude and of Elevation above the Level of the Sea (p. 156); The Icy Region (p. 158); the Region of Moss (p. 158); the Region of Barley and Northern Agriculture (p. 159); Marine Vegetation within the Arctic Circle (p. 160); Vegetation on the Snow (p. 163); Terrestrial Vegetation of the Far North (p. 164); and of the Forest Zone (p. 174).

SECTION II.—*Forests,* - - - - - 176
Forest Trees and Forest Products of the Government of Archangel (p. 176); and of Olonetz and Vologda (p. 187); Details of the Appearance and Contents of the Forest Estate of Vuig (p. 179).

SECTION III.—*Classified List of Plants found in the vicinity of Lake Onega by Forst-Meister A. Guenther,* - - - - - - - 182

SECTION IV.—*Vegetation in Lapland,* - - 191

SECTION V.—*Palaeontological Botany,* - - 193
Views advanced by Dr Oswald Heer, and expounded and illustrated by Count Saporta, relative to Vegetation having originated in the Far North, and diffused itself Southwards.

CHAPTER III.—*Fauna,* - - - - - 236

  SECTION I.—*Quadrupeds,* - - - - 236

  SECTION II.—*Birds,* - - - - - 242

  SECTION III.—*Insects,* - - - - - 245
Notice of Insects injurious to Forest Trees in Northern Russia by Forst-Meister Guenther (p. 245).

    SUB-SECTION A.—List of Coleoptera collected by Mr Guenther in the Government of Olonetz, arranged according to *Catalogus Coleopterorum Europae et Caucasi Auctoribus,* L. V. Heyden, E. Reitler, et J. Weise (p. 248).

SUB-SECTION B.—List of Lepidoptera collected by Mr. Guenther in the Government of Olonetz, arranged according to *Catalog der Lepidopteren des Europaeischen Tannengebietz*, Von O. Standinger, u. M. Wocke (p. 264).

---o---

AUTHORITIES CITED.

ACERBI, p. 157; *The Arctic World*, pp. 76, 160; BALFOUR, p. 231; *Blackwood's Magazine*, p. 73; BARON V. BUCH, p. 156; BUFFON, p. 199; DAA, p. 152; HEPWORTH DIXON, pp. 49, 69, 126; DUEDEN, p. 147; *Edinburgh Encyclopaedia*, pp. 83, 149; FORBES, p. 233; *Forests of Finland*, pp. 13, 85; *Frost and Fire*, p. 32; GUENTHER, pp. 182, 245, 248, 264; GUILLEMARD, p. 60; HAYES, p. 167; HEER, pp. 193, 204; HOWISON, p. 189; *Hydrology of South Africa*, p. 89; *Journal of Forestry*, p. 108; JUDRAE, pp. 36, 44, 86, 96, 122, 133, 137, 176; LEONARD, p. 94; LONGFELLOW, p. 18; LYELL, p. 224; MORDVINOFF, p. 158; NICOLSON, p. 13; NORDENSKJOELD, pp. 198, 213; PEARS, p. 94; *Pramiatnais Knjka*, p. 43; RAE, p. 130; *Russian Songs*, p. 56; SAPORTA, pp. 195, 208, 209, &c.; *Scotsman*, p. 233; *Ustaff Laesnoi*, p. 123; LADY VERNAY, p. 95; WAHLENBERG, pp. 59, 62, 191; MACKENZIE WALLACE, p. 155; WEREKHA, pp. 101, 122.

FOREST LANDS AND FORESTRY

OF

NORTHERN RUSSIA.

---o---

PART I.

FOREST LANDS.

INTRODUCTION.

IN the introduction to a companion volume on *The Forest Lands and Forestry of Finland* it is stated that 'I spent the summer of 1879 in St. Petersburg, ministering in the British and American Chapel in that city, while the pastor sought relaxation for a few months at home. I was for years the minister of the congregation worshipping there; and I had subsequently repeatedly spent the summer among them in similar circumstances. I was at the time studying the forestry of Europe; and I availed myself of opportunities afforded by my journey thither through Norway, Sweden, and Finland, by my stay in Russia, and by my return through Germany and France, to collect information bearing upon the enquiries in which I was engaged. On my return to Scotland I contributed to the *Journal of Forestry* a series of papers, which were afterwards reprinted and published under the title *Glances at the Forests of Northern Europe*. In the preface to this pamphlet I stated that in Denmark may be studied the remains of forests in prehistoric times; in Norway, luxuriant forests managed by each proprietor as seemeth good in his own eyes; in Sweden, sustained systematic endeavours to regulate the management of forests in accordance with the latest deliverances of modern science; in Finland, *Sartage* disappearing before the most advanced forest economy; and in Russia, *Jardinage* in the north, merging into more scientific management in Central Russia, and *reboisément* in the south.'

The following pages may be considered a study of information I then collected, together with information which I previously possessed, or have subsequently obtained, in regard to the forestry of the Russian Governments of Olonetz, Vologda, and Archangel, through some of the forest lands of which I made a tour in the summer of 1882.

CHAPTER I.

THE NEVA.

THE steamer plying between St. Petersburg and Lake Onega takes its departure from a quay nearly opposite to the Finnish Railway Terminus. Of the passage by water from the centre of the city to this, I have given an account in the companion volume, entitled *Forests and Forestry in Finland.* The drive by land from the centre of the city to the quay of the Onega steamer may be less striking, but it is not less interesting.

Starting from Vassiliostroff, or from the English Quay, passing along this brings us upon the Isaac's Plain, now the Alexandra Sadd. This was the scene of the military insurrection which occurred in December 1825, on Nicolas I. succeeding to the throne. I write from memory of what was told to me fifty years ago by men who had seen, and men who had acted in the conflict, and of what I then read of the trial and condemnation of leaders in the fight, and the visions which rise before me may be more vivid than absolutely correct, but they are my remembrances accurately given. The conspiracy had been progressing rapidly during the later years of Alexander I. His death, and the succession of the Grand Duke Constantine, intensified the desire of many to effect a change in the government of the Empire. By a family compact Constantine had ceded to his younger brother Nicolas all claim to the throne. There, as here, the Sovereign never dies. The oath of allegiance to Constantine had been taken when the death of Alexander was proclaimed; and now the soldiery were required to take an oath of allegiance to Nicolas. The disaffected officers, assuming an air of loyalty, said they could not unless personally released from their previous oath by the Grand Duke to whom they had sworn fealty.

According to the accounts given to me, arrangements were made for a general insurrection of all the troops in the vicinity of St. Petersburg. The day was fixed for their concentrating on St. Petersburg; something occurred to disturb that arrangement, and to render expedient a short postponment of the insurrection; but one regiment did not receive notice of this in proper time, and began their march; the others on hearing this were precipitately mustered and marched from their cazernes, and mustered in the Isaac's Plain.

The Emperor Nicolas, it is said, came out from his palace to the troops with his son, the late Emperor Alexander II., then a child, in his arms; and addressing the soldiery in a dignified tone, he demanded of them: 'What is it you want? Is it to take my life, and the life of my son? If so, we are here; but what will taking our lives benefit you, or advance your purpose? What better will be your condition; No, no, my children, stand by your Tsar.' The effect was, as might be imagined, electric; and enthusiasm found expression in shouts of loyal attachment to the Tsar. But while some cried 'Long live the Emperor,' by others were raised the cry, *Constantine e Constitutzio.*

My informant happened to be on the Isaac's Plain when one regiment entered from one direction, and another from another. From their bearing he concluded something unusual was going on, though what it might be he could not imagine. Wishing to see what might be seen he took up his station along with others between pillars in the wall of the Senate House. When he saw what was occurring fain would he have got to his home; but this seemed impracticable; at length a gun was so planted as to sweep the whole Galernoy—the street at his feet; and then he made away with all speed and at all risk, glad enough to get safely to his house. From him I learned not a little of what occurred before the bloodshed commenced, and the spot was indicated to me by him, where, standing between the two pillars of the Senate House, whence he was looking down upon the commotion, he saw an officer ride up to two or three soldiers, who were standing at his feet, and greatly excited, and cry to the men, 'Call out *Constantine e Constitutzio!*' The men hesitated, grounded their arms, and insisted to know who *Constitutzio* was. 'Constantine's wife, you blockheads,' was the reply. '*Ah Xoroshos!*' (all right), and forthwith the cry was raised, *Constantine e Constitutzio!*

While these shouts were being raised a shot was heard, whether fired accidentally or of design was unknown, and Milardovitch, the Military Governor-General of the city, who was riding in front of the troops addressing to them soothing words, fell dead. He was a man universally beloved; great confusion immediately ensued, and fighting began amongst those who had marched thither animated by a common purpose. Soon regiment was firing upon regiment, which of them in the supposed interests of the Crown, which of them in the supposed prosecution of rebellion, it would have been difficult to say. Dreadful was the slaughter. At length one party remained masters of the field—either they were the loyal party, or if they were not, and there were none such in the fight, they found it convenient to proclaim themselves such; and order was re-established. It was in winter, the snow on the plain was everywhere red with blood; but during the night which followed, openings were made in the ice which covered the Neva, which flows past the place; and into these, it is said, the dead and the dying were thrown promiscuously. These the rapid-flowing river carried quickly away; the snow was cleared off, and similarly disposed of; and by the following day peace was restored.

I said to one of the officers of a so-called loyal regiment: 'Now, tell me, were you all perfectly loyal?' 'As loyal as man is to God,' was his reply, given with great solemnity and assurance. I said, 'That is not in accordance with what I have heard from others.' His rejoinder was: 'Mark my reply; I cannot say more, but, I repeat, As loyal as man is to God.' 'Oh! ho! Universal depravity?' He said nothing in reply, but his silence led me to say: 'How came you then to fire upon those who were in the same plot?' He shrugged his shoulders and replied, 'We are in the army accustomed to implicit and immediate obedience. Our superior officers gave the orders to us; and we in turn, faithful to our traditions, gave them to those who were under our command, *et Voila!*'

Through the scene of this outbreak, and slaughter, and treason, and treachery, lies the way from the English Quay to the point of embarking for Lake Onega.

St. Isaac's Plain is now laid out as a garden with public promenades, and there there is to be seen one manifestation of the reign of peace which, often as I have beheld it, always gives me pleasure. Here and there are laid down cart-loads of fine clean sand in the summer, in which children, whose parents cannot arrange to take them to the country, as do most who can, may work as they please—making dove-cots, digging pits, raising bulwarks, and trundling sand from one corner to another, as children delight to do; and in the early morning, before they are again astir, all is swept up again into a heap, where during the day they or others may resume their play.

Within this garden stands the statue of Peter the Great, on its immense boulder support. At right angles to the Synod and Senate Houses stands St. Isaac's Church, the dome of which dominates the city. Along the left-hand side of the garden are the Admiralty Buildings, with their golden spire, opposite to which diverge at equal angles the three lengthened Prospects which divide into sections a great extent of the city situated on the mainland,—one of them, the Nevsky Prospect, being one of the celebrated streets of Europe. Beyond the Admiralty Buildings is the Imperial residence, the Winter Palace, looking out upon the monolith erected to the memory of Alexander I., and on the

Glavno Stab. Beyond the palace, and connected with it, is the Hermitage, containing a valuable collection of articles pertaining to Peter the Great, the founder of the city, and an invaluable collection of paintings by ancient and modern artists, of coins, of cameos, and of other gems. A little way brings us upon the Champs de Iars, an extensive plain devoted to reviews of the troops, dominated by palaces, among others that erected by the Emperor Paul, and in which he met his death. In front are the Summer Gardens, studded with statuary. Leaving the Champs de Mars, we pass a statue of Kotussof, and passing in front of the Summer Gardens we pass a shrine for prayer, erected on the spot where was made the first attempt to assassinate the late Emperor Alexander II. Passing onwards between a noble quay or line of palatial residences and the river, and passing the entrance to a noble granite bridge spaning the Neva, and leading to the Finnish railway, and the country beyond, we reach at length the quay from which the steamers for Lake Onega take their departure.

The commencement of the voyage is through miles of urban scenes—houses, churches, manufactories, and wharfs; but these past, the rural scenery is reached. Here the banks of the Neva present aspects differing greatly from those of the Saima Canal in Finland: there the banks are wooded to the water's edge, approximating and receding, and branching off into numerous lakelets, and presenting in front ofttimes a wooded barrier against advance, which, however, is found practicable by some narrow outlet in a concealed corner; here there is a broad expanse of river, winding indeed, but never so as to conceal what is ahead. Both banks of the Neva, from St. Petersburg upward, for a considerable distance, are crowded with timber yards and manufactories of different kinds, and not until Alexandrof, eight or nine miles distant by road, has been left a considerable way behind is it otherwise. Beyond this the banks are studded with villages, with 'datches' or villas, and with churches admirably located for effect, occupying prominent positions in the landscapes seen from a considerable distance as they are approached by the river.

There is not a lack of trees; but these do not constitute a characteristic feature of the scenery. On the north bank of the river, the land a mile or more in breadth, has been sold or ceded to private parties, Russian communes, German colonies, and Finnish villagers, but beyond this is a forest, belonging the Imperial Domains, 160 square versts in extent, preserved for the chase, where bears, wolves, elks, blackcock, capercailzie, and ptarmigan constitute the principal game. While this forest on the right bank, commencing a little distance from the river, but not seen from it in general, has been preserved, and may be said to extend almost continuously from the Finnish frontier to the Ural Mountains and Siberia, with what was once a forest, I may say, in continuation of this on the opposite bank of the Neva, it is otherwise.

Along the left bank of the Neva, which at no distant period was richly wooded, the woods have been extensively destroyed, sometimes by forest fires, sometimes otherwise. The agricultural operations adopted on both banks have in many cases, perhaps in most, been the following: The ground has been cleared of the stumps and roots, which, after being piled and thus dried, have been used as firewood; the ground then roughly ploughed, and, though all hillocks and hollows, has been sown with oats or rye, generally the former, and a remunerative crop, though not abundant, has been obtained. The stubble has then been ploughed in, and the ground in steep furrows exposed to the influence of the weather. In early spring it has been again ploughed, harrowed, and levelled; and potatoes, planted with appropriate manure. For two or three years thereafter oats, barley, or rye, are grown, but the rye, not being suitable for malting, can only be used in the manufacture of pearl barley, for which there is no great demand; with the last crop, the field is laid down in Timothy grass and wild clover, and for some years hay crops are taken. The red clover gradually gives place to white clover, which grows abundantly where woods have been burnt, and the Timothy grass gives place to some extent to other grasses less nutritive to horses and cattle, but which still yield a valuable hay. After a time the same routine is repeated.

On some farms the ground is divided into four, five, or six sections, each of which in succession is planted with potatoes, with appropriate manure, or sown with grain. Winter-sown rye yields a beautiful crop, but the risks from early frost are so great as to frighten many from adopting this method of culture.

When the winter-grown rye makes what is deemed too great progress, it is eaten down with cattle or mown, by which operations the number of stoles is probably increased. If in spring the crop threatens to fail, it is generally ploughed up, and the ground left in fallow.

Though damage is done to the hay by rain, a copious rainfall immediately before cutting the grass is hailed with delight as greatly facilitating the work of the mower. The mowing is generally done during the night, throughout which there is abundant light in July in this region. The cut grass is turned and tossed by women the following day, and by nightfall or next morning it is fit for stocking in hay-cocks.

The German colonists give more special attention to the growth of potatoes, and only introduce the other cultures in so far as this can be subordinated to the successful growth of the potato.

Four hours' steaming brings the traveller from St. Petersburg to Schlusselburg, the fortress of that name being situated on an island in the river, the town on the shores of Lake Ladoga, from which the river takes its rise. The Neva has a course of about 40 miles, the medium breadth of its main stream is about 1500 feet, and the depths of its mid channel, near St. Petersburg, is about 50 feet.

The population of Schlusselburg numbers about 3,500. The island on which the fort is built is between 300 and 400 yards long. The walls of the fort are about 50 feet high, of great thickness, and fortified in the old fashion, with turrets and battlements. The passage to and from the mainland is by a drawbridge. So early as 1324, George, Prince of Moscow and Novogorod, built a fort here while on an expedition against Wyborg, which was taken by the Lithuanians, who in turn were driven out by Magnus, King of Sweden, A.D. 1347, but it was retaken by the Novogorodians in 1352. It was ultimately, in 1702, occupied by Peter the Great, but till that time it was a subject of frequent contention between Russia and Sweden. Since then the fortress has often served as a State prison; here one Emperor at least, John VI., met his death, and here the first wife of Peter the Great was confined after being divorced.

CHAPTER II.

LAKE LADOGA.

LAKE LADOGA is the terminal reservoir of the waters drained off from Finland by the Saima See and the Falls of Imatra, and the reservoir of waters drained off by other water-courses and water systems in the north, and the east, and the south, where all these waters are collected, to be thence discharged by the Neva, and conveyed by it to the Gulf of Finland and the Baltic, and thence by the Katigat and the Skagar Rack into the German Ocean and the Atlantic beyond.

Lake Ladoga is the largest lake in Europe: its length from north to south is 138 miles, and its greatest breadth 90 miles, the area of the lake is 6,300 square miles. It contains several islands, and numerous rocks and sand-banks, which render the navigation of it dangerous. It is fed by about sixty tributary streams, the principal of which are the Volkhov and Siasi on the south, and the Svir, which connects it with Lake Onega in the Government of Olonetz. The dangerous character of the lake, and the frequency and violence of its storms, induced Peter the Great to begin the formation of a canal from Schlusselburg to Novaia Ladoga, on the Volkhov, which was completed in 1732. Additional canals to extend the means of communication were dug under the direction of Catherine II. The Ladoga Canal, 70 miles in length, and 74 feet in breadth, forms with the Siasi and Svir canals, a continuous line round the south and south-east sides of the lake.

The name Novaia Ladoga, or New Ladoga, was given to the town mentioned, a town of 3,000 inhabitants, in contradistinction to Staraia Ladoga, or Old Ladoga, formerly a large place, noted in Russian annals as the earliest residence of Ruric, first sovereign of Russia. The ruins of its ancient walls are still seen; but since the erection of Novaia Ladoga, it has become almost depopulated, and the number of its houses does not exceed fifty. Steamers ply daily during summer between St. Petersburg and Novaia Ladoga, leaving Schlusselburg at 2.30 P.M., and arriving at Novaia Ladoga at 1 A.M., and thence there is water connection with the Volga and the network of inland navigation.

The creation of this canal, skirting the southern extremity of Lake Ladoga, was one of the last enterprises of Peter the Great. In 1718 he formed the plan of the canal and its sluices. He relied much on canals as a means of extending the inland navigation of his country; and when canals were to be dug with this view, sometimes on marshy and almost impassable grounds, he was frequently seen at the head of his workmen, digging the earth, and carrying it away himself.

The purpose he had in view in the formation of this canal was to establish a communication between the Neva and another navigable river for the more easy conveyance of merchandise to Petersburg, without making the circuit of the coast of the lake, which, from the storms which prevailed on the coast, was frequently impassable for barks or small vessels, and still is, even for steamers, as I have experienced in crossing it. The Emperor levelled the ground himself, and the tools and implements used by him in digging up and carrying off the earth have been carefully preserved. His courtiers followed his example, and persistently prosecuted the work, which at the same time they looked upon as part of an impracticable undertaking. It was not finished till after his death; but at length it was completed.

The object was to draw produce to St. Petersburg for exportation; and in extension of the project of the Ladoga Canal he, in the same year, or shortly thereafter, made another canal by which the Caspian became connected by navigable water channels with the Gulf of Finland and the Ocean; and boats sailing up the Volga, traversing a canal connecting this river with another, proceeding so far by this and by another canal to the lake of Ilmen, could thence by the Ladoga Canal reach the Neva, whence goods and merchandise might be conveyed by sea to all parts of the world.

Lake Ilmen, in the Government of Novogorod, intersecting a town of the same name, is connected with Lake Ladoga by the Volkhoff. The length of the course of this river is about 150 miles. It is deep and rapid, but except when its waters are low, when it forms cascades, it is navigable. It is connected by canal with the Siasi, which flows through the Government of St. Petersburg in a N.N.W. direction, throughout a course of about 100 miles.

Schlusselburg forms thus a port of departure whence the traveller may proceed by water to the south, to the east, to the north, or to the west. By the canal the traveller may proceed by water to Odessa, the Black Sea, Constantinople, the Mediterranean, and thence whithersoever he will, the wide world over. By leaving the Volga, a little beyond Kazan, and ascending the Kama to Perm, a railway journey of 312 miles will bring him to Ekaterineburg, in Siberia, which is in like manner possessed of wonderful facilities for inland navigation.

In a sketch of the Hydrography of Finland, in a volume entitled *The Forest Lands of Finland*, I have narrated the experience of my friend, the Rev. W. Nicolson, agent of the British and Foreign Bible Society, in descending the Ulea River to the Gulf of Bothnia, whence he found his way by coasting steamers to St. Petersburg. It was by this route that he had entered Finland. Embarking at

St. Petersburg, and ascending the Neva at Schlusselburg, he entered on Lake Ladoga, and visited Konevits, Hexholm, and Walamo: the first and the last of which are holy islands, visited by hosts of pilgrims by which the steamer was crowded, most being deck passengers. Many of these were very sick, and strewn on the deck, it seemed as if in many places they were lying three deep. A day or half a day spent by the steamer at each place gave an opportunity to such as chose to fulfil there the object of their pilgrimage, and proceed with the steamer to the next; but if they preferred to do so, they could remain, and proceed by the following steamer. At one of the islands, which is covered with forest, no tree may be felled, no animal slaughtered. Other restrictions similar in character, but different in kind it may be, are in force at other holy places subject to the Greek Church of Russia; and yet———. Well, at Konevits, a somewhat lugubrious-looking priest took up his position by the side of the vessel, and never left it while the vessel was in port. My friend asked the captain what was the purpose of his there keeping watch and ward, and he was informed that the priest was stationed there by ecclesiastical authority to watch, and afterwards to testify, that no one of the monks had obtained brandy on board or from on board the steamer: a precaution which may be commendable; but one which may naturally suggest a gibe at the expense of the character of the consecrated men; and allegations to their prejudice in regard to sensual immoralities are advanced in support of the gibe.

He entered Finland by Hexholm, a little town beautifully situated on the western shore of the lake—a town of which frequent mention is made in the old historical annals of Finland. On the shores of Lake Ladoga, with its forest-crowned hills and lovely valleys, one may feel as if transported to some one of the most lovely regions of the South.

My route took me across Lake Ladoga in an easterly direction to the mouth of the Svir, by which are conveyed to this gigantic reservoir the waters of Lake Onega, which is in its turn largely fed by the upper waters in the mountain range which constitutes the boundary between Russia and the north-eastern läns of Finland.

CHAPTER III.

THE SVIR.

THE aspect of the shores of the Neva, seen from the river, differs not more from that of the Saima Canal than does the appearance of Lake Ladoga differ from that of the Saima See, the latter studded with islands, or branching out in innumerable lakelets—thus one broad expanse of waters like to Lake Ontario, and some of the other lakes of North America, presenting nothing to view within the horizon but water, water, water, generally smooth, but liable like these to be tossed into billows, when lashed by a storm.

But near the mouth of the Svir wooded islands again appear, and ere the voyager is aware, he has passed from creeks between islands into the continuous flow of the river, which, in consequence of its rapid descent, has a current of considerable force. The surface of the river throughout lengthened stretches present appearances characteristic of rapids—now that of the surface of molten lead, now that of a shallower stream passing over a rocky bed, and at times the steamer quivers as it stems the torrent. The banks are covered with trees, but most of them comparatively young. Floating rafts of timber, barges laden with deals, piles of firewood a fathom in height, stretching in some cases as at Vajnee for versts along the shore, tell of what has occasioned this. All the older trees have been felled, and these are the reproduced forests in a condition which may be compared to that of youth and early manhood.

Four hours or four and a half hours brings the steamer across the lower end of the lake, from Schlusselburg to

Sermaksi on the Svir, where are large stores for produce and a meteorological observatory. Some two hours or more brings it to Ladonoi Pole, founded by Peter the Great, and formerly a naval dockyard. Here there are still extensive bakeries, to which flour is brought from great distances, and whence are shipped great quantities of *cringles*, a kind of Russian or Swedish biscuit, made in the form of a long roll, the tapering ends of which are twisted together so as to form a ring with an expansion in the middle. Great quantities of cray fish are caught in the neighbourhood, and offered for sale by peasants crowding the landing stages where the steamer touches. It is situated at what appears to be the confluence of two rivers.

Ladonoi Pole (the field of Lodi) is a place of some interest, being the spot where Peter the Great built his first galleys in 1702. He superintended their building in person, and subsequently employed them in taking the fortress of Schlusselburg from the Swedes. A monument in cast iron marks the site of a house in which Peter resided.

In four hours or more is reached Vajnee, where apparently the rafts of timber are made up. This is brought hither in floats made in three tiers of twenty logs each, bound firmly together. Here ten such are connected in a long line, two oar-like helms or helm-like steering oars are attached to each end of the long raft, and either end may become stem or stern, or alternately the one or other. Ten or twelve women, with one man amongst them to direct their movements, ply those on the foremost float, so as to keep the whole in the current, or to move it out of the way of steamers advancing in an opposite direction, and one or two men do the like with those in the stern. The women whom I saw thus employed were cleanly dressed, and looked healthy and strong, neither coarse-featured nor inelegant in form. Throughout the whole region women are extensively employed in rowing the boats plying on the river, and also in piling firewood.

Such work is preferred to domestic service, and is more highly remunerated, and I found complaints rife of difficulty experienced in procuring domestic servants, especially during the summer months, the wages paid being two roubles, or 4s a month!

There were many villages besides those I have mentioned, and the whole country presented the appearance of an older settled district than those of Finland through which I had passed. There was much more land under pasturage and agriculture. Contrary to what is seen in some parts of Russia, all the houses and outbuildings stood erect and in good order. One striking feature was the clean, newly-painted appearance of all the churches. Here, as elsewhere, there are little erections containing pictures of the saints, before which the passing traveller may offer his prayers. Some of these were ruinous, though with most it was otherwise, and, with the churches, all were beautifully clean, leaving no occasion for another Longfellow to tell of what the Devil saw in Church.* The people, with a slight admixture of Russians from the south, seemed generally to be of the Thutchi tribe, and those further to the north to be Karells.

---

* What a darksome and dismal place!
I wonder that any man has the face
To call such a hole the house of the Lord,
And the gate of Heaven—yet such is the word.
Ceiling, and walls, and windows old,
Covererd'd with cobwebs, blacken'd with mould;
Dust on the pulpit, dust on the stairs,
Dust on the benches, and stalls, and chairs.
The pulpit, from which such ponderous sermons
Have fallen down on the brains of the Germans,
With about as much real edification
As if a great Bible, bound in lead,
Had fallen, and struck them on the head;
And I ought to remember that sensation!
Here stands the holy water stoup!
Holy water it may be to many,
But to me the veriest Liquor Gehennae!
It smells like a filthy fast-day soup!
Near it stands the box for the poor,
With its iron padlock, safe and sure.
I and the priest of the parish know
Whither all these charities go;
Therefore, to keep up the institution,
I will add my little contribution!
  (*He puts in money.*)

The flow of the Svir is W.S.W. Its course is about 150 miles. Its principal affluents are the Ivina and Vagena, flowing into it on the right bank, and the Oiat and Pacha on the left.

The Ivina rises some 25 miles S.S.W. of Petrozavodsk, the capital of the Government of Olonetz, and has a course of about 60 miles. The Oiat rises in the same Government, and flowing westward enters the Svir after a course of 92 miles. The Pacha rises near Ledia, in the Government of Novogorod, and flowing first west, and then north, joins the Svir after a course of 150 miles. Its principal affluent is the Kapcha.

The Canal of Siasko connects the Svir with the Polkhov, and thus forms a means of communication between St. Petersburg and the surrounding provinces.

Russians express their delight in heat in a proverbial saying that 'Heat breaks no bones!' and in sweltering weather officers and others in like position are to be seen on the streets of St. Petersburg in wadded cloaks and overcoats, and peasants in sheepskin shoubs; but it is not without cause, for changes in temperature are great and sudden. Spaniards have a proverbial saying that 'The zephyr which will not extinguish a candle may blow out a man's life,' and another to the effect, 'Sit in a draught and send at once for a lawyer and a priest, to make your will and receive your dying confession.' A similar opinion seems to prevail in Russia, and of this I had an illustration in the course of my voyage. I was on the upper or steering deck; the steerage passengers covered their deck, sleeping in all attitudes and places, and the cabin passengers were seated or walking about on theirs, when all at once, like a picture of the resurrection from the dead, the steerage passengers started to their feet, and men and women alike were in movement, like the sea in a storm, putting on their shoubs, and the cabin passengers in continuous lines were making for the cabin doors as if at the summons of a church bell. I was about to ask the occasion, when I became aware that the wind had suddenly veered round to the north-east, and was blowing somewhat strongly. This was the occasion of the sudden movement!

In making arrangements for another journey, I asked a friend, who had travelled extensively in the region I was purposing to visit, what provision of clothing for the journey I should make? He said, 'Go where you may in Russia, always provide for four different temperatures, otherwise you are not safe.' It was said playfully; but on this trip one day we had the temperature of 92° Fahr., next day that of 67°, and the day following 42°. Calling the attention of one of my fellow travellers to this, he said that in Archangel, where he resided, one day they had a temperature of upwards of 90° Fahr., and in the course of a few hours it was frost! One day while on this trip I felt the heat extreme, but within twenty-four hours the cold was such that I could not sleep at night though wearing my under flannels, and covered with a pile of coverlets.

I met also on this trip with an incident illustrative of the feelings with which my countrymen are regarded by the Russians. In Russia fellow travellers freely enter into conversation with one another. There are sufficient indications of their position in society to prevent unpleasantness; and brotherly kindness is one of the traits of character seen alike in prince and peasant. There was on board the steamer a gentleman, an official in the Forest Service, between whom and myself there sprung up considerable intimacy and freedom of conversational intercourse, from our both being interested in forestry and in several allied matters. On the second or third day he said to me, laughingly: 'I must tell you this: When I came on board, the captain said to me, "There is an English tourist on board; he will be ignorant of our language; he is going to the Government of Olonetz; and as you also are going there, I wish you would give to him any assistance he may need in travelling." I at once said, "No; he is an Englishman. I know not but any advance made by me may bring upon me an insult, the English are so supercilious. I will have nothing to do with him." I find that you are the tourist; I find nothing supercilious about you. How is this?' I replied, also laughingly, 'I am not an Englishman.' 'You are not an Englishman? I thought you were.' 'Oh, no.' 'Then what countryman are you?' 'I am a Scotsman.' 'Ah,' he exclaimed, 'that explains all;' and with fervour he embraced me, giving me, as is the national custom, three kisses—the first on one cheek, the second on the other, and the third on the first again.

I told some friends in St. Petersburg of the incident, when my story was capped, with other like incidents experienced by others; and it was mentioned by one whose experience had been given, that gentlemen in Russia fully recognise the difference between Scotchmen and Englishmen. They say the average Englishman is a Jingo, pooh-poohs anything you may say, and will not hear you complete a sentence you may have begun; the average Scotchman is intelligent; he is not afraid to hear what you have got to say; he may differ from you, but he will allow you at least to express your views, and he will judge dispassionately of what you say. I have found the difference between Scotchmen and Englishmen recognised the wide world over, and generally with a preference for my countrymen. To many foreigners the supercilious bearing of Englishmen is offensive; and of English-speaking people the only thing more so is that of a discourteous American citizen travelling *au prince*.

CHAPTER IV.

LAKE ONEGA.

At the issue of the Svir from Lake Onega is Vosnisenya, one of the principal centres of inland navigation by a widely extended system of canals, of which there are three connecting the Baltic and North Sea with the Volga and the Caspian. One of these commences here, where are collected barges and vessels from all parts of Russia, including an extensive region of Siberia. In this respect it resembles the town and port of Schlusselburg on Lake Ladoga, it being the fort from which that town takes its name, and not the town itself, which is situated on an island in the river.

In the month of June last (1883) were formally opened at Serumaxa on the Svir two new canals, connected with the rivers Svir and Siass, the formal opening taking place in the presence of the Emperor and Empress, who were accompanied by several Ministers. The Svir Canal has been named after the Emperor, and the Siass Canal after the Empress. Both canals are 8 feet deep, and will allow of the passage of large vessels, thus rendering possible the transport of goods to the harbours of St. Petersburg in ten days less time than hitherto.

Vosnisenya is called by the inhabitants The St. Petersburg Gate. I received much kind courtesy here, and I am indebted for much information and assistance in my enquires to Forst-Meister Dlŭtofsky. Here I had an opportunity of penetrating a little way into the forests, along with Forst-Meister Herman Goebel, from whom I received much information in regard to forestry in Russia, and more especially in regard to planting operations on the steppes in the vicinity of Odessa. The trees which I saw were chiefly the pine (*pinus sylvestris*), and the fir (*abies excelsor*); but from another forest official, from whom also I derived much valuable information, I learned that in the forests beyond there were also to be found the Norwegian maple, the lime, the elm, the juniper, and other kinds of trees, the juniper attaining to an arborescent size, very different from the juniper bush of Britain.

Finding it difficult to thread our way through the close growing trees, I asked how a forester found his way when lost in the wood. The reply was, 'The coniferæ are on the north side of the trunk more or less densely covered with lichens, and thus we know in what direction to go.'

Lake Onega may be considered a basin of the Vodla, the principal river flowing into it, while the outlet is by the Svir. Its water is clear, and abounds in fish. The bases of the islands, of which it contains several, are limestone. It is by the Vodla and Mariienskai water-course that it is connected with the Volga, while by the Svir it is connected with Lake Ladoga, the Neva, and the Baltic, into which river flows also the Oiat, which rises in the Government of Olonetz, and, entering that of St. Petersburg, it joins the Svir on the left bank, after a course of 92 miles. The Vodla flows from a lake bearing the same name, 26 miles long from north to south, and 14 miles in breadth, to the N.N.E. of Pudoj, or Pudoscha, a town with a population of 1200 inhabitants, situated about 65 miles east of Petrozavodsk. Flowing first in a S.S.E., and then in a S.S.W. direction, it falls into Lake Onega after a course of about 100 miles.

From Vosnisenya I sailed to Petrozavodsk. Lake Onega measures about 220 versts, about 150 miles in length, and about 75 versts, or 50 miles, in breadth. Petrozavodsk is situated on the western shore of the lake. The town dates from 1701, when Peter the Great established works there for casting cannon. These were afterwards destroyed, and replaced by other works completed in 1774. Guns continued, nevertheless, to be imported into Russia at great expense from the Carron works in Scotland, owing probably to the unsatisfactory state of the establishment on Lake Onega. In order to improve the latter, Catherine II. invited Charles Gascoigne, the manager of the Carron works, to come over and rebuild the Gun Foundry, which he did in 1794, when the town that had sprung up around it took the name of Petrozavodsk. Gascoigne was accompanied by two English artisans, who subsequently rose to great eminence in the service of Russia. Guns for the navy are to this day cast at Petrozavodsk.

The Government is divided into seven districts, and the town of Olonetz, on the river Olonza, is the capital of the district bearing that name, while Petrozavodsk is the capital of the Government. Olonetz is a town of nearly 3000 inhabitants, on the Olonka, at the junction of the Megvega, about 15 miles from the east coast of Lake Ladoga, 132 miles north-east from St. Petersburg, and 72 miles south-west of Petrozavodsk. It has large building docks, established by Peter the Great, and numerous saw-mills. The soil, where not covered with forests, is in some parts stony, and in others marshy, and generally little capable of culture. Between Lakes Onega and Ladoga are quarries of marble and porphory, and in some of the mountains are mines of iron and copper. In the Museum at Petrozavodsk are beautiful collections of the marbles found in the Government, and models of different places of interest in the Government, and relics of the Imperial founder.

Petrozavodsk covers a good deal of ground, surmounted by two cathedrals, in both of which officiates the Archbishop of the diocese. Near to these is the residence of the Governor, and the Cazerne or Barracks. The town is traversed by a small river, the Lossalenka. The number of inhabitants is about 7000.

Lake Onega, upon which it has been built, is only one of several lakes in the Government of Olonetz.

The Olonetz chain of mountains, on the confines of the Russian Government of the same name and of Finland, constitute part of the watershed whence the waters flow on the one side into the Baltic, and on the other into the White Sea. In continuation of this chain on the north-west are the mountains of Maanselkä, extending from Finland to Uleaborg, at the head of the Gulf of Bothnia; and again, on the north-west of these, they are connected with the Dofrines or Dovre-field, a name sometimes given to the whole Scandinavian mountain system, but more explicitly in application to that portion which, in latitude 62°, 63°. N., extends from Cape Stadtnaes to the Sylt-Field, or Syll-Fiellen, in Norway, throughout its length, dividing the bason of the Baltic from that of the White Sea.

The Government of Olonetz, bounded on the west by the Grand Duchy of Finland and Lake Ladoga, is bounded on the north and north-east by the Government of Archangel, on the south-east by that of Vologda, on the south by that of Novogorod, and on the south-west by that of St. Petersburg; it lies between 60° and 64° 30′ N. lat., and 29° 40′ and 41° 40′ E. long., measuring 390 miles in length from N.W. to S.E. and about 300 miles at its greatest breadth, with an area of 51,100 square miles. With the exception of the range of hills on its north-west boundary, the surface of the Government is generally level, but interspersed with undulating hills. It comprises districts forming portions of the basins of three far-separated seas—the White Sea, the Baltic, and the Caspian. In the first-mentioned, the north and east of the Government, is Lake Latcha, in which the Onega river and Lakes Sego and Viga have their sources, and in which are numerous sheets of water of smaller dimensions; in the second are Lakes Onega and Ladoga, the principal tributaries of which are the Vodla and the Vitegra; and in the third is the Kovja.

Lake Latcha is about 24 miles in length from north to south, and 8 in breadth. It receives the waters of the Soid; and gives origin to the river Onega, flowing to the

White Sea. Lake Sego is 30 miles in length from N.W. to S.E., and 24 in breadth. To the north-east of this is Lake Vigo, fed by the Vig, which enters it on the south-east, and flows from it on the N.N.W. The Kovja takes its rise in Lake Kovjskoe, in the southern part of this Government; flowing south, it enters the Government of Novogorod, and falls into Lake Bielo, on its N.W. side, after a course of 60 miles.

In the number of its lakes, and the relative proportions of land and water, Olonetz resembles Finland; but the area of the land greatly preponderates over that of the lakes, and in waterfalls and rapids the similarity of the two countries is maintained, though the similarity is not very great.

CHAPTER V.

THE FALLS OF KEEWASH.

I AM indebted greatly to the Forst-Meister in charge at Vosnisenya, and not less so to the Forst-Meister in charge at Petrazavodsk, and to the Oberforst-Meister Günther, whom I met on my return voyage. Desirous of seeing something of the forests beyond, with the advice of the Forst-Meister in charge, and accompanied by his brother, also a forest official, I proceeded from Petrozavodsk to the Falls of Keewash, which took me through some stretches of old forest, as well as extensive stretches of forests in a state of rejuvenescence, and land which, reclaimed from the forests, had been devoted to agriculture. The latter showed a fertility which justifies those who, though lamenting the inconsiderate destruction of wood, tell that the forests are not to last for ever, and that even the destruction of them may be made the means of promoting the advancement of a country. My excursions into the forests took me over well nigh a hundred miles, and were deemed sufficient to give me a general idea of the condition of those existing in the district. The road which we took brought us in sight of some beautiful lakes, sprinkled with beautiful islets, generally wooded to the water's edge.

I had here an opportunity of seeing one of the *Objestchicks*, or Forest Circuit Wardens, in his home. This was anything but a palace. It consisted of but a single apartment, with a projection—I cannot call it a verandah—extending the whole breadth of the house, and some ten feet deep. My fellow traveller and I arrived at midnight, and the wife was immediately in attendance to make arrangements for our comfort. In this verandah were all kinds of agricultural implements, made of wood alone, a two-pronged and a three-pronged fork, cut from the branching bough of a tree, and a harrow with smaller branches inserted into the bars for teeth. The place was lumbered with other encumbrances. On a wooden bedstead, under a covering like a mosquito net, but made of coarse linen, the man lay awake. His wife showed us to noble apartments in a wood built pavilion, erected by the Government in connection with the *zavod* already mentioned, and designed for the accommodation of visitors, bringing as we did a permit from the authorities. It was light, for at midsummer there is no night there, but it was very cold, and we returned to the house, where, according to our desire, she was preparing for us a supper of tea, eggs, black bread and butter. Inside there was like confusion; but there was a plain deal table and bench, scrupulously clean. Behind a temporary screen two daughters were sleeping; on the hearth was a blazing fire, on which our eggs were being cooked; of these there was being prepared no stinted supply, and as soon as they were boiled they were placed for a minute in cold water. Meanwhile, with embers from the fire, the water in the *samovar*, or Russian tea-urn, was made to boil, and tea was soon infused. We intimated our preference to take it there rather than in the pavilion. Bread and butter were soon on the table, but no knife or spoon. My fellow traveller at once produced his pocket-knife, and laid on thickly the butter; seeing I had none, our hostess soon produced her husband's forester's knife with which the bread had been cut, and handed it to me, when I did likewise. Our only light was that supplied by the little fire on which the eggs were boiled, and a small window not above 18 inches square, and though it was now only half an hour past midnight the light was all we could desire. There was no chimney, but two holes in the roof, about a foot square, one above the fire, the other near the centre of the room. The higher half of the apartment was filled with smoke, which irritated my eyes, and provoked a cough, upon which our hostess brought out a coverlet and spread it on the floor for me to sit upon, while my fellow traveller and she, to use an Aberdeen expression, *newsed away* about anything and everything.

We, warmed and refreshed, retired for the night to the pavilion, where was a spacious sitting-room, with bow-window commanding the Fall, plainly, but elegantly and substantially furnished. There were more than one bedroom furnished in like style, a dressing-room with every thing pertaining to the toilet, and a small cabinet with everything pertaining to the writing-table, and outside was a kitchen with hot plates and other conveniences, but there was no bed or table linen, knives or forks, or tea or dinner crockery. All these visitors were expected, in accordance with the usage of the country, to bring with them, together with provisions, unless they chose, as did we, to procure these from the woman in charge.

The night was cold, and in the morning we were again fain to betake ourselves to the house of the *Objestchick* for our morning meal, rather than have it served in our elegant quarters. On going there we found the daughters, as well as the parents, all astir, the former making up for sale small bundles of birch twigs, which are used extensively throughout Russia for switching the body in the national bath. The husband, who retained his bed when we arrived at night, was now up and ready for conversation. He had five watchmen under him, and an extensive district under his charge, and he appeared to talk intelligently of much that related to his forest duties, and of much beside. His wife had the oven charged with wood in full blaze, and this added not a little to our comfort in the chilly morning. Tea and eggs, black bread and butter, were served to us *ad libitum*, the butter being laid on thick, and the eggs drunken out of the shell. The smoke, as volumes came belching out from the open oven, was still more offensive to my eyes than that of the evening before. As then, a coverlet was spread for me on the ground, and a pillow placed upon the bench upon which to rest my arm supporting my head, while I listened to the conversation carried on by my fellow traveller and our host. It appeared that on the preceding night, a bear had entered the kraal, and was hugging one of the cows in a death-gripe, preparatory to carrying it off, when it was disturbed by the bell of our *tarantass*, and the rattling of our vehicle over the planks of the bridge by which we approached the pavilion. Bears are abundant in the forests there. Twenty carcasses of cows hugged to death by bears had been found in the neighbourhood of Petrozavodsk in the course of the preceding year, and game of all kinds abound. Winged game, rabschick and tytark, may be had for a mere trifle, and, the mystery of hare-soup being still unknown, hares are a drug, the people being unwilling to eat them, and when they do prepare them for the table they are larded with strips of bacon introduced with a larding pin before being cooked, otherwise, they say, the flesh would be too dry to be eatable.

The wages of this forest warder were 18 roubles, or 36s, a month, with eight dezatines, or 20 acres of arable land, free pasturage for his cows, and the horse which he is required to keep. He and his wife had been there for eighteen years. By way of 'stirrup cup' our hostess had prepared for us a cup of tea at our hour of starting, and I having asked for a glass of milk, it seemed to give her greater pleasure than even it did to me, to give me as much milk as I chose to drink.

I may mention that other forest officials are remunerated in the same way—salary, dwelling-house, and arable land, varying with their rank and position. They hold rank corresponding to that in the army, with corresponding uniforms of which the higher officials have four sets—one appropriated to work in the forests, another to work in the office, another so-called full dress uniform, and a fourth characterised as undress uniform. On retiring from the service they retire with rank next higher in grade to that which they held, and with permission to wear the corresponding uniform; but not until they have attained to the rank of General have they any claim to a retiring allowance or pension.

From the pavilion is seen a magnificent view of the falls; and there has been constructed below the fall a footbridge, more than half a verst long, leading towards a little wooden temple on the higher level, from which the most striking view of the falls may be had, and other views are obtained in passing along this bridge.

The Falls of Keewash are on a river by which a higher-lying lake within the Russian boundary empties its waters into a series of lakelets by which they find their way into Lake Onega, and thence by the Svir into Lake Ladoga. The Russians distinguish between rapids and a waterfall; the latter they call *koski*, the former *koskia*. The Falls of Imatra may be cited as a specimen of the *koskia*. The Falls of Keewash are, strictly speaking, a specimen of the *koski*. As the Falls of Niagara are, divided by Goat Island into two distinct waterfalls, so it is with Keewash: from the right bank of the river, not the left, as in Niagara, there is a miniature resemblance of the Horse-shoe Fall, and for a little way behind the surface of the upper stream may be seen from the shore the vacant space over which the water shoots; but soon this is broken into what I can only describe as a gigantic counterpart to the falling of the laps of the wig of the Speaker of the House of Commons, and that worn by the Lord Chancellor of England.

Beyond the dividing island there flows away the remainder of the stream, but by far the greater portion of this makes its escape by the side, pouring over and between ridges of rock like the teeth of a comb, and forming a continuation of the fall. A small portion makes its way behind the pavilion to the lower basin.

Elsewhere I had seen logs dashing over waterfalls— rushing along 'seething, boiling, tumbling, racing waters,' and had looked down upon the basin into which the waters fell, 'in whose circling depths logs and tree-trunks, stripped of bark and water-worn, swept round and round, and anon raised a despairing arm to heaven for help, only to sink back into the toils again.'

Of such a scene the author of *Frost and Fire* gives the following graphic sketch. It is an account of what was seen by him at Vigelund, on the Torristal River, about ten miles above Christiansand.

'At every moment some new arrival comes sailing down the rapids, pitches over the fall, and dives into a foaming ground pool, where hundreds of other logs are revolving and whirling about each other in creamy froth. The new comer first takes a header, and dives into some unknown depth, but presently he shoots up in the midst of the pool, rolls over and over, and shakes himself till he finds his level, and then he joins the dance. There is first a slow sober glissade eastward across the stream to a rock which bears the mark of many a hard blow. There is a shuffle, a concussion, and a retreat, followed by a pirouette sunwise, and a sidelong sweep northwards up stream towards the fall. Then comes a vehement whirling over and over, or if a tree gets his head under the fall, there is a somersault, like a performance in the Halling dance. That is followed by a rush sideways and westward, when there is a long fit of setting to partners under the lee of a big rock; then comes a simultaneous rush southwards, towards the rapid which leads to the sea, and some logs escape and depart, but the rest appear to be seized with some freak, and away they all slide eastwards again across the stream to have another bout with the old battered pudding-stone rock below the sawmill; and so for hours and days logs whirl one way, in this case against the sun, below the fall, and they dash against the rounded walls of the pool. Such is the effect of these concussions that above the fall it has been found necessary to protect the rock against floating bodies so as to preserve the way of the stream. It threatened to alter its course and leave the mill dry, for the rock was wearing rapidly. Lower down, nearer the sea, is a long flat marsh, between high, rounded cliffs; and there these mountaineers, floating on to be sawn up, form themselves into a solemn funeral procession which extends for miles; and it may be noticed that the course of this stream of floats is always longer than the course of the river's bed; for the water is slowly swinging from side to side as it flows, and the floats show the course of the stream and its whirling eddies.'

It was from the banks of the river at the side of the fall that I got my best view of the cataract. Immediately above the fall lay moored a long raft of logs ready to be shot. We were informed of this before leaving Petrozavodsk, and a hope was expressed that we might see it done, but in this we were disapponted; and it was a disappointment, for this is always an exciting scene. I had visited the locality described in the passage cited, and here was everything combined to produce a similar scene—the waterfall and the basin below. I had, however, to rest satisfied with imagining what the scene would have been.

I have found few things in connection with forestry more exciting than incidents connected with the flotage of timber.

On the Glommen, in Sweden, I have seen hundreds and thousands of logs floating down the river separately, to be collected and arranged according to the owners' marks upon them at a depôt at a lower level. The breadth of the river, compared with the size of these logs, suggested the idea of some boys having emptied into a brook a hundred or a thousand boxes of matches, and of these being floating away. At any little fall of three or four feet, there they came tumbling down, sometimes sideways, sometimes slanting, and sometimes head foremost, and kicking up their heels in the air. Occasionally in some of the rivers in Norway the trees floated thus accumulate, and become so interlaced that further progress is impossible.

There, as elsewhere, logs are transported from the spot where they are felled to the banks of the nearest stream, and marked with the initials of the owner. On the melt-

D

ing of the ice they are pushed into the current, and the contributions of many affluents find their way to the river, which may at the time be covered with the floating masses, which become more or less compactly interlaced, till some projecting rock in the bank or the river bed arresting some, others are impeded and stopped in their course, and ultimately many thousands, it may be, are stopped, and piled up in a confused heap. It is perilous work to break up the piled mass, and set the logs afloat upon the stream again. Elsewhere 'the men employed go about balancing themselves on detached logs in the middle of the stream, pushing on each log by means of a boat-hook, till at last the mass of logs hanging together begins to be disturbed and shake, and then comes the struggle for the men to regain the shore. The skill which the men display in disentangling the logs, the agility with which they run about and maintain their balance on the floating logs, as well as on those which are fixed, the intelligence which they apply to the separation and setting afloat again of all those interlaced logs, and, in fine, the courage with which they face all these perils, are all of them worthy of admiration.' The statement is cited from a report by Dr. Brock, a distinguished Norwegian statistician.

The author of a large work entitled *Frost and Fire*, to which I am indebted for the account of logs performing the Halling dance below the waterfall on the Torristal river, some distance above Christiansand, tells that after the logs have been launched 'many get waterlogged and sink; and these may be seen strewed in hundreds upon the bottom, far down in clear green lakes,' and he goes on to say:—

'Many get stranded on the mountain gorges, and span the torrent like bridges; others get planted like masts amongst the boulders; others sail into quiet bays, and rest upon soft mud.

'But in spring, when the floods are up, another class of woodmen follow the logs and drive on the lingerers. They launch the bridges, and masts, and stranded rafts, help them through the lakes, and push them into the stream;

and so from every twig on the branching river floats gather as the river gathers on its way to the sea.

'Sometimes great piles of timber get stranded, jammed, and entangled upon a shallow, near the head of a narrow rapid; and then it is no easy or safe employment to start them. Men armed with axes, levers, and long slender boat-hooks, start down in crazy boats, and clamber over slippery stones and rocks to the float, where they wade and crawl about amongst the trees, to the danger of life and limb. They work with might and main at the base of the stack, hacking, dragging, and pushing, till the whole mound gives way, and rolls and slides rumbling and crashing into the torrent, where it scatters and rushes onwards.

'It is a sight worth seeing. The brown shoal of trees rush like living things into the white water, and charge full tilt, end on, straight at the first curve in the bank. There is a hard bump and a vehement jostle; for there are no crews to paddle and steer these floats. The dashing sound of raging water is varied by the deep musical notes of the battle between wood and stone. Water pushes wood, tree urges tree, till logs turn over and whirl round, and rise up out of the water, and sometimes even snap and splinter like dry reeds.

'The rock is broken, and crushed, and dinted at the water-line by a whole fleet of battering-rams, and the square ends of logs are rounded; so both combatants retain marks of the strife.'

At Keewash I was told that the logs shot the fall bound together in floats or rafts, whether single floats consisting of 60 logs, or in long rafts consisting of ten such floats, bound together, I neglected to enquire, but I presume the former. I would have been glad to have seen the effect in either case, but I could not await the operation.

CHAPTER VI.

FORESTS OF OLONETZ.

WISHING to learn a great deal more in regard to the general appearance of the forest lands in Northern Russia than could be obtained on such a holiday trip as I could myself undertake, I asked Professor Schavranoff, Director of the *Laesnoi Corpus*, or School of Forestry in the vicinity of St. Petersburg, how this could be accomplished. He at once supplied me with a narrative prepared by M. Judrae, a forest official of high position, of a tour of inspection which was made by him in 1867. The following is a translation of part of his narrative of what he saw :—

'The first steamer of the season (1867) proceeding from St. Petersburg to Petrozavodsk, sailed on the 30th May (Old Style), having been prevented from sailing earlier by the ice on the Neva and Lake Ladoga. With fine, somewhat warm weather, we left the capital, and a few hours' hard steaming against the current brought us to Lake Ladoga; but scarcely had we got 30 versts (20 miles) from St. Petersburg when ice began to meet us, some of it in sheets of a very large size; and it was getting dark. The keen north-east wind made itself felt; and looking to the horizon there stretched out before us a sea of unbroken or of congealed fields of ice; the steamer, however, resolutely advanced. I took refuge in the cabin from the intolerable cold, but after a few minutes I hastened on deck in consequence of the steamer being stopped. There was ice in immense shoals ahead of us, so that to go on in the course we were following would have risked damage to our paddle-wheels, whereby we should have been placed in an awkward condition amongst the ice floes of the Ladoga. At length the order was given to cast the anchor and wait for the day. In a few minutes we were fast, and a strangely contrasting stillness and silence pervaded the vessel, while a magnificent scene was stretching around us in all direction. Far as the eye could see were open spaces of water and sheets of ice commingled, and whole schools of black seals moving backward and forward on the floating masses, while with the cold wind were combined black clouds and a murky sky, although it was now the 31st of May (O.S.), the 12th day of June in lands where the New Style has been introduced.

'Next day the steamer by some way or another got through Lake Ladoga, and entered the river Svir. Steaming along, we found everywhere on the banks on both sides, woods, woods, woods. From the deck of the vessels could only be noticed firs, and pines, and birches, although in some parts of the Government of Olonetz there still grew the Norway maple, the lime, the elm, and other kinds of trees.

'Now we passed on the left bank of the river the town of Ladenoi-Pole, founded by Peter the Great, and formerly a naval dockyard. A few hours more and we reached Vosnesenya, one of the principal centres of inland navigation by a system of canals, of which there are two or three connecting the Volga with the Baltic.

'The village of Vosnesenya is situated on the Svir as it issues from the Lake Onega, and it is called by the inhabitants the Petersburg Gate.

'It was impracticable to go further by the steamer, as the ice in this lake had not yet broken up; consequently I had to travel to Petrozavodsk by horse, which I did by a very picturesque route by the western shore. From Vosnesenya to Petrozavodsk by the so-called Vilegarskoi road is 130 versts, or 86 miles.

'Between the hills are occasionally met with rivers or rivulets flowing into Lake Onega. The current of these is very rapid in consequence of the steep declivity of the ground towards the lake; and they present generally the characteristics of mountain streams. The most striking feature of the country is the great quantity of boulders upon its surface, the number of which, if stated, would be almost incredible. They consist exclusively of granite and other primary or transition formations, covered partially by drift, in which is a red sand in considerable quantity. There are also projecting from the ground granite hills in whole or in part quite bare, or covered only with lichens. Having examined the works I proceeded further. On the right hand was a magnificent view of the Onega Lake, the breadth of which at this place is above 80 versts (about 54 miles). On the left side of the road were hills, the continuation of those of Finland, which pass into the Government of Olonetz, but fall away towards the south till they present an altitude not exceeding 420 feet above the level of the country around. The surface of the hills seen from this point is covered with forests, which consist of four different kinds of trees, intermixed in varying proportions—fir, pine, birch, and aspen. The height of these trees, judging by the eye, seemed to be low compared with like vegetable productions. The only impression I have retained of the course of the whole journey to Petrozavodsk, with such opportunities of observation as I had, was a feeling that I had gotten into a comparatively northern region, and that I must be nearing the polar circle; granite hills and interminable forests, a stony soil, with abundance of waters but a sparse population—these are my remembrances of my first acquaintance with the Government of Olonetz.

'Every twenty or thirty versts (14 and 20 miles) there were small villages inhabited by Karrells, a tribe of Finns who have retained the Finnish language, but in every other respect they are like the population from Novgorod found in the south and east, and in parts of the central portion of this government.

'Within two weeks after my arrival at Petrozavodsk I was once more on the road in my *kibitka* speeding onward to the most northern town in the Government of Olonetz, where, according to the opinion among the population, is the end of the world. This town is called Povonetz. At about 20 versts, or 14 miles, from Petrozavodsk is the village of Thouya, the first post station on the river of the same name, across which there is a barge ferry. The river Thouya flows into the Onega Lake, and has throughout its course a very rapid current. Where I crossed there was wood being floated down from the Government mining forest estates situated further up, from whence the strength of the current brought them down.

'The current brought them with such rapidity and force that the barge was in danger, and with difficulty we reached the other shore.

'The rapid current is not favourable for the flotage of timber, and there has been formed what may be called a dam at the mouth of the river; but this having been broken, a great quantity of wood has been carried into the Onega Lake, whereby the navigation of it in this part by steamers has been impeded. It is to be desired that some effective measures were taken to prevent this loss, which increases the cost of what forest timber is secured.

'Looking at the floating timber I was struck with the activity with which the men employed maintained their footing, each standing on a log and holding in his hand a long pole or boat-hook, with which he balanced himself, and with which, in floating down the timber, he cleared the obstacles encountered; and these on this river are very numerous.

'For this purpose it is generally inhabitants of the district who are employed, these being very skilful and accustomed to the work. They are here known as "Onejan," or Onega men, and I am under an impression that under this general name such workmen pass in St. Petersburg.

'Proceeding onward to the north, on both sides of the road there were to be seen forests and forests, and nothing but forests. I can affirm that the person who is acquainted with the extent of these forests only by knowing the number of *desatins* which they cover, has no idea of what that extent is. To obtain this one must travel through them—travelling continuously through forests for five hundred versts; and he must experience personally the depressing influence produced by the forests and forest-covered mountains of this forest region to enable him even partially to comprehend what is implied in the easily pronounced statement about so many millions of *desatins*. Such numerical statements are required for the production of a *national tax*, or estimate and description of what fellings should be made to secure a sustained production of wood, and the charge to be made for trees; and the latter is a matter which is not so easy of accomplishment as to many at first sight it may appear to be. Those who are in the trade do not make known what is the cost of preparing the timber for the market, or the prices obtained by them, being afraid of the charge to them being raised. If there be made but a simple allusion to the subject, they begin to complain that they are carrying on their operations at a loss, and that the demand for timber is diminishing from year to year. And to arrive at a knowledge of the truth, the forest officials must solve the problem for themselves, with such data as they have at command. . . . . .

'At a distance of ninety versts, or sixty miles, from Petrozavodsk, is the village of Leejma, where there is a saw-mill of considerable magnitude, occupied also at the present time by M. Baelaeff. It is erected on the river Leejma, and has two water-wheels and four frames of saws, two for each water-wheel. It works without intermission day and night, and can cut up in the course of the year 60,000 logs; but, in consequence of hindering circumstances, it cuts up only some 45,000. These are pine logs of the length of twenty-two feet, and eight verschocks or fourteen inches thick at the upper extremity. The boards most in demand in the market are twenty-two feet long and three inches thick, which are known as $2\frac{1}{2}$-in. boards; and besides these there are what are called inch boards, sent chiefly to Holland. According to the statements of the traders these inch boards are both in quality and price inferior to the Swedish boards of the same measurement, in consequence of which the preparation of them in large quantities is not remunerative.

'Coming next to those connected with Povonetz, I have to state that not far from the post road on the river Koumsa, at a distance of twenty-three versts from Povonetz, there is a saw-mill belonging to the timber merchant, Mr Zachanieff. This mill also I had an opportunity of seeing. It is built in a very pretty situation, in the valley of the rapid river Kamsa, surrounded by lofty hills extending to the Onega Lake. The mill has one wheel and two frames, and there are sawn in the course of the year about 30,000 logs. Everywhere about it are seen order and cleanliness; and there is a fire which never dies out, burning continuously the outside slabs, the ends of logs, and other *débris;* and what are literally mountains of sawdust fill up the picture of the mill and its surroundings, while the noise of the wheel and of the saws is reverberated by the surrounding forest.

'A journey of some fifteen miles brings us to Povonetz. A poorer and more unattractive town than this it is impossible to imagine: it is simply a village built on the plan of a town. The most remarkable object in Povonetz is an old wooden church standing on the shore of Lake Onega, built by Peter the Great, the only monument which indicates that ever he was here. There is, it is true, besides this, the Petrozavodsk road; but this is now only a footpath or track, by which are brought the goods obtained in this town from Archangel. Add to this two or three legends or traditions about Peter, and all records of his having been here are exhausted.

'Almost close to the town, on the estuary of the Povetchanka, is the saw-mill, which gives some little life to the town, and is the only thing which vivifies its existence.

'The whole biographies of the place tell only of what relate to the works, besides which the inhabitants have an opportunity several times in the course of the summer to admire a steamboat which visits the place; but beyond this and fishing, change they have none.

'Almost all the vessels which leave the landing-place of Povonetz are laden with boards produced at this mill. In the fullest sense of the word, Povonetz is a timber town, and on arriving here I felt proud while I thought that my profession was the principal profession of its inhabitants, and had to do with the very source of its wealth. To determine and specify what is the trade of the place must occasion no difficulty to any one. Its imports consist of everything excepting wood and fish, and its exports consist of wood and fish alone, the latter principally Triska.

'The discharge of my professional duties led me further in the north.

'For nine versts or six miles beyond Povonetz it is possible to travel by wheel, but beyond this point the journey has to be made by water in very uncomfortable boats on narrow lakes and rivers connected with them. From the Lake Volozer issues the river Povetchanka, which flows through a very picturesque country. Thanks to the high hilly shores, the general rapid current of the river, and the frequent occurrence of considerable rapids, this little river, or rivulet, is in spring changed into a very dangerous torrent, tearing along, and threatening to engulf and carry along with it whatever may tumble into its waters. It has a course of about eleven versts, nearly eight miles, and by it are floated some 20,000 logs a year to the saw-mill at Polonetz.

'The construction of a road from near the Lake Volozer to the White Sea has been projected, and the initiative of the execution has been taken, but nothing more seems to have been done. The proposal created great excitement throughout the district, where there are very few roads of any kind or other facilities for communication with other parts. Scarcely could the projection of a railroad in any other part of Russia produce so much discussion, and excite so many hopes, as would the making of a common road in this country. This part of the Government of Olonetz is passing through that period of its history at which any measures taken for the formation of roads, the opening up or clearing of forests, or the introduction of regular systematic agriculture, possess very great interest.

'Unhappily the execution of this enterprise has not proceeded further than the felling of a strip of trees through the forest along which it was proposed that the road should be made. And the general impression is that soon the whole matter will end, for money is not forthcoming, and the kind of road is not satisfactory. Coming upon it at various points, it seemed to me that the projector or surveyor had of design made it to pass at a distance from the most important centres, and carried it over uninhabited districts and unsuitable land.

'For forest operations this road to the White Sea would not have been unimportant, and, having referred to the subject, I am led to mention also a proposal which has been made to open up water communication between the White Sea and the Onega Lake. Having no accurate data, but only partial information, I cannot give details or discuss fully the importance of this gigantic subject.

'Of this proposal it is stated in the *Pramiatnais Knjka* or official Notes of the Government of Olonetz for the year 1867, " The execution of this project, opening up communication between the White Sea and the Gulf of Finland, and *vice versa*, proposed solely with a view to commercial enterprise, would for strategical purposes affecting the whole of the north of Russia have immense importance;" and Mr Seederoff [a gentleman well known throughout this region, a merchant who has carried on great commercial transactions in Archangel and Nova Zembla, and made valued contributions to the different International Exhibitions in the capitals of Europe] says in a communication to the Imperial Free Economical Society, "Steam war vessels could proceed from Cronstadt and make their appearance for the protection of the inhabitants of the shores of the White Sea, or, if necessary, of Archangel, which now, in consequence of the dismantling of the fortress of Nova Dwina, is left without defence."

'According to the views of Mr Seederoff, there will only be required the construction of a canal fifty versts long, which, opening on the lake, will make it possible for shipping to pass from the lake to the White Sea, or from the White Sea to Lake Onega, and, consequently, to St. Petersburg.'

M. Judrae goes on to say, 'Mr Seederoff has, I think, neglected to take into account the rapids of the Svir, which, to the accomplishment of such a scheme, would require to be passed by a canal; and this would add considerably to the difficulty of the undertaking. But both the Onega and White Sea canal and the White Sea road remain at present within the category of projects, and they are likely to remain there for some time, as no one seriously believes in the execution of either of them in the immediate future.

'Returning to details of my journey : After proceeding some eighteen versts, or twelve miles, by boat through a succession of narrow lakes, I landed at a place where there was a very narrow path, which could only be traversed on foot. A walk of six versts, or four miles, brought me to the village of Morskoy Mosselgie. The road I found pleasant. It goes along a picturesque ridge of hills, running from west to east some thirty-two versts or twenty-one miles north of Povonetz, at an elevation of some seven hundred feet above the level of the adjacent country, being the greatest altitude in the Government of Olonetz.

'This ridge constitutes the watershed of streams flowing on the one side to the Baltic, and on the other to the White Sea. On the former are narrow lakes, which, with the rivers connecting them or issuing from them, flow into the Onega, while on the latter is the Matkozero, whose waters flowing northward follow the course indicated.

'On the banks of the Matkozero they fell wood for the saw-mills at Povonetz, transporting it by carts across the Mosselgie ridge, the woodmen going further and further into the interior of the forest, in consequence of the exhaustion of the woods near to the saw-mill.

'Having crossed the ridge, I found myself in a country manifesting all the characteristics of a northern land. I got into a boat again, and went by the river some ten versts to the village Telekin situated on a river or lake of the same name—I say river or lake because it is difficult sometimes to designate precisely what is seen by the one name or the other, or to tell at what point it ceases to be one or the other, and to take the different character where it should be called a narrow lake and where it should be designated a broad river.

'The general character of the waters in these regions is the following:—Picture to yourself a comparatively small lake, having a flow barely noticeable in some one direction. In the direction of this flow the water becomes perceptibly narrower, and the shores get higher, and the water takes the form of a river, distinguishable from the lake above by being narrower and having a greater current, or it becomes a strong rapid, by which the waters flow into a large expanded lake, which serves as a reservoir for the waters of the surrounding neighbourhood. Such are the general characteristics of all the small expanses of water in this region.

'All the rivers and rivulets here have a great many rapids throughout their course. For example, the river Vuigozero, which in a course of 100 versts, 66 miles, from its leaving the lake of that name to its flow into the White Sea, has seventeen rapids. The fall of the river through these rapids is 272 feet. In consequence of these rapids all navigation of the river is out of the question. Only timber is floated down these rivers and their confluents in spring—and this notwithstanding the stones with which the beds are filled, and other obstacles. From the Vuigozero Lake I went 40 versts, 27 miles, by the river Telekin to its embouchure.

'The Vuigozero or Vuigor Lake is one of the largest lakes in the district of Povonetz. It is 60 versts or 40 miles long, and 30 versts or 20 miles broad. It is throughout its whole extent studded with islands, which, according to the idea prevailing in the locality, are equal in number to the days of the year. Some of these have an area of 50 square versts. Many are covered with woods, but uninhabited and unsurveyed, so that their contents are unknown; many of them find no place on the map, and their area is considered as lake, though some of them have good available soil, or are covered with valuable forests of pine.

'On the shore of the Vuigozero is the Vuigozero Podost, the most southern station in the Government of Olonetz for village administration, and this uninviting spot must be for a time my place of residence.'

Mr Judrae states what his duties were, and communicates some valuable information relative to forest operations there and in similar localities, all of which may be afterwards given in detail. Here it is this journey and the aspect and condition of the country as seen by him which alone engages our attention. Of this he thus resumes details:—

'My duties on the forest estate of Vuig being finished, on the 9th of September I left this place to go further to the north and the north-east, to that part of the Povonetz district inhabited by the Corrells or Karrells; I had to go by boat on the Vuigozero. We had a favourable wind, and the well-filled sails carried the small boat along with great rapidity.

'I might have proceeded directly to my destination, but I could not deny myself the pleasure, being there, of visiting the village of Voitzi, situated at the northern extremity of Vuigozero, 100 versts distant from the White Sea, and in the Government of Archangel. This village is well known as the site of a gold mine, which is now a thing of the past. Gold was discovered there in 1735 by a peasant Tarass Antonoff. Mr Poushkaroff, in describing the Government of Archangel, says that at Voitzi, quartz on being crushed and washed yields 7½ zolotnicks of gold for every 150 pounds.*

'The working of the mine at Voitzi was discontinued in 1783. In 1827 gold was discovered on the banks of the river Vuig, and in the course of the present century investigations have been made several times by private parties, but they have not proved successful in unearthing any stores of the precious metal, and at the present time there remain only here and there pits and buildings in which the workmen of a former day were lodged. The traditions of the district give in a thousand different forms pictures of the prosperity enjoyed by the peasants in those times. I had to pay somewhat dearly for the gratification of my curiosity to see the old mine. I had to go on foot 35 versts, 22 miles, to the nearest Karrell village, situated on the edge of the forest estate of Padan; and my walk was the more unpleasant because the road, or, speaking more correctly the path, according to local phraseology, founded on the topographical condition of the ground, lay *across the earth*, and did not go *with* the *earth*. From the northern part of the district of Povonetz, on to the shores of the White Sea, the ground lies in parallel rows of ridges or linear hillocks, with hollows consisting sometimes of peat bog lying between. The ridges are narrow and long, as are likewise the bogs by which they are separated. They run north and south, consequently for the traveller in either of these directions the path lies along the summit of the ridge, and according to local phrase he goes *with the earth*, and it is more easy to do so; but if he travels east or west he must walk across the bogs and ridges; and as the crests are about a verst apart, this has to be done in every verst of his journey.'

'The Padan forest estate lies to the west of Vuigozero, and covers an area of 570,000 *desatins*.† All that has been

---

* 96 zolotnicks = 1 lb. Russia, 40 lbs. Russ. or 36 lbs. avordupois = 1 pood.—J.C.B.
† A *desatin* = 40 × 60, or 2,400 Russian square fathoms of 7 English feet. An English acre is 0·37041 of a *desatin*, which makes a *desatin* = 2·69972 English acres. A *desatin* = 4·2789 Prussian morgens. A verst is equal to two-thirds of an English mile.—J. C. B.

said of the forests of Vuig might be reaffirmed of these, the only difference being that these show a decided preponderance of pines over the number of firs, especially in the northern parts. In general, the nearer we approach the sea the more rarely do we meet with fir, and at last this tree disappears entirely. With regard to the quality of the pine I can state as the result of my personal observation, that within certain limits the quality of the wood does not depend on the latitude, but is in direct relation to the quality of the soil. In the Vuig forests they are met with first in the middle and northern parts of the Padan forest estate.

'With the development of forest operations in this district the Padan forest will acquire much importance in consequence of the number of navigable streams existing in the White Sea basin. In the southern part of this forest estate there is the Segozera, second in size only to the Onega; further to the north is the Lake Ondazero, through which flows the river Onda, one of the tributaries or confluents of the Vuig, constituting the boundary between the Governments of Olonetz and Archangel.

'On the 1st of October I crossed the frontier of the Government of Olonetz, and in three days I was on the shores of the White Sea.'

CHAPTER VII.

FORESTS OF ARCHANGEL.

THE report of M. Judræ does not embrace any account of the forests in Archangel. Of these, and of the aspects of forests along a different route from that followed by him, some idea may be formed from the graphic accounts of Hepworth Dixon, in the volume entitled *Free Russia*. His journey was from north to south—from Archangel towards the central districts of the empire—and it thus supplies an account of what might have been seen on a return journey from Archangel or from beyond it by another route. It is in accordance with what I have myself seen travelling in other parts of Russia, and with what I have heard from others of what they have experienced in travelling through the forest lands of the Empire.

Speaking of his tour through Russia, he says :—' My line from the Arctic Sea to the southern slopes of the Ural Range ; from the Straits of Yeni-Kale to the Gulf of Riga ; runs over land and lake, forest and fen, hill and steppe. My means of travel are those of the country ; drojki, cart, barge, tarantass, steamer, sledge, and train. The first stage of my journey from north to south is from Solovetsk to Archangel ; made in the provision boat, under the eyes of Natha John. This stage is easy, the grouping picturesque, the weather good, and the voyage accomplished in the allotted time. The second stage is from Archangel to Vietegra ; done by posting in five or six days and nights ; a drive of 800 versts through one vast forest of birch and pine.'

It is the narrative of this journey to which I have referred as conveying some idea of what travelling in forest lands in Russia is, and of what is seen on such a journey. His conveyance was a tarantass, which he thus describes:—'A tarantass is a better sort of cart, with the addition of splash-board, hood, and step. It has no springs; for a carriage slung on steel could not be sent through these desert wastes. A spring might snap; and a broken coach, some thirty or forty miles from the nearest hamlet, is a vehicle in which very few people would like to trust their feet. A good coach is a sight to see; but a good coach implies a smooth road, with a blacksmith's forge at every turn. A man with roubles in his purse can do many things; but a man with a million roubles in his purse could not venture to drive through forest and steppe in a carriage which no one in the country could repair.

'A tarantass lies lightly on a raft of poles; merely lengths of green pine cut down and trimmed with a peasant's axe, and lashed on the axles of two pairs of wheels, some nine or ten feet apart. The body is an empty shell, into which you drop your trunks and traps, and then fill up with hay and straw. A leather blind and apron to match keep out a little of the rain; not much; for the drifts and squalls defy all effort to shut them out. The thing is light and airy, needing no skill to make and mend. A pole may split as you jolt along; you stop on the forest skirt, cut down a pine, smooth off the leaves and twigs; and there, you have another pole! All damage is repaired in half-an-hour.'

A tarantass was supplied to him for the journey by a private friend, and the British Consul supplied him with a trustworthy servant to do what was needful by the way, and fetch back the vehicle when the journey was completed; and the journey is thus described:—

'This private tarantass is brought round to the gate; an empty shell, into which they toss our luggage, first the hard pieces—hat-box, gun-case, trunk; then piles of hay to fill up chinks and holes, and wisps of straw to bind the mass; on all which they lay your bedding, coats, and skins. A woodman's axe, a coil of rope, a ball of string, a bag of nails, a pot of grease, a basket of bread and wine, a joint of roast-beef, a teapot, and a case of cigars, are afterwards coaxed into the nooks and crannies of the shell.

'Starting at dusk, so as to reach the ferry at which we are to cross the river by daybreak, we splash the mud and grind the planks of Archangel beneath our hoofs. "Good-bye! Look out for wolves! Take care of brigands! Good-bye! good-bye!" shout a dozen voices; and then that friendly and frozen city is left behind.

'All night under murky stars we tear along a dreary path; pines on our right, pines on our left, and pines in our front. We bump through a village, waking up houseless dogs; we reach a ferry, and pass the river on a raft; we grind over stones and sand; we tug through slush and bog; all night, all day; all night again, and after that all day, winding through the maze of forest leaves, now turned and scared, and swirled on every blast which blows. Each day of our drive is like its fellow. A clearing thirty yards wide runs out before us for a thousand versts, the pines are all alike, the birches all alike. The villages are still more like each other than the trees. Our only change is in the track itself, which passes from sand drifts to slimy beds, from grassy fields to rolling logs. In a thousand versts we count a hundred versts of log-road, two hundred versts of sand, three hundred versts of grass, four hundred versts of waterway and marsh.

. . . .

'If the sands are bad the logs are worse. One night we spend in a kind of protest, dreaming that our luggage has been badly packed, and that on daylight coming it shall be laid in some easier way. The trunk calls loudly for a change. My seat by day, my bed by night, this box has a leading part in our little play; but no adjustment of the other traps, no stuffing in of hay and straw, no coaxing of the furs and skins, suffice to appease the fitful spirit of that trunk. It slips and jerks beneath me, rising in pain at every plunge. Coaxing it with skins is useless; soothing it with wisps of straw is vain. We tie it with bands and belts; but nothing will induce it to lie down. How can we blame it? Trunks have rights as well as men; they claim a proper place to lie in; and my poor box has just been tossed into this tarantass, and told to lie quiet on logs and stones.

'Still more fitful than this trunk are the lumber vertebræ in my spine. They hate this jolting day and night; they have been jerked out of their sockets, pounded into dust, and churned into curds. But then these mutineers are under more control than the trunk; and when they begin to murmur seriously I still them in a moment by hints of taking them a drive through Bitter Creek.

'But, ah! here is Holmogory. Holmogory was the birthplace of Lomonosoff, a philosopher and a poet of the last century, whom his countrymen greatly honour—— here is Holmogory, standing on a bluff above the river, pretty and bright, with her golden crops, her grassy roads, her pink and white houses, her boats on the water, and her stretches of yellow sand; a village with open spaces; here a church, there a cloister, gay with gilt and paint, and shanties of a better class than you see in such small country towns; and forests of birch and pine around her —Holmogory looks the very spot on which a poet of the people might be born.

'From Holmogory to Kargopol, from Kargopol to Vietegra, we pass through an empire of villages, not a single place on a road four hundred miles in length that could by any form of courtesy be called a town. The track runs on and on, now winding by the river bank, now eating its way through the forest growths; but always flowing, as it were, in one thin line from north to south; ferrying deep rivers, dragging through shingle, slime, and peat; crashing over broken rock; and crawling up gentle heights. His horses four abreast, and lashed to the tarantass with ropes and chains, the driver tears along the road as though he were racing with his Chert—his Evil One; and all in the hope of getting from his thankless fare an extra cup of tea.

It is the joke of a Russian jarvey, that he will "drive you out of your senses for ten kopecs." From dawn to sunset, day by day, it is one long race through bogs and pines. The landscape shows no dykes, no hedges, and no gates; no signs that tell of a person owning the land. We whisk by a log fire and a group of tramps, who flash upon us with a sullen greeting, some of them starting to their feet. "What are those fellows, Dimitri?" "They seem to be some of the Runaways." "Runaways! Who are the Runaways? What are they running away from?" "Queer fellows, who don't like work, who won't obey orders, who never rest in one place. You find them about here in the woods everywhere. They are savages. In Kargopol you can learn about them."

'At the town of Kargopol, on the river Onega, in the province of Olonetz, I hear something of these Runaways, as of a troublesome and dangerous set of men, bad in themselves, and still worse as a sign. I hear of them afterwards in Novogorod the Great, and in Kazan. The community is widely spread. Tinvashef is aware that these unsocial bodies exist in the provinces of Yaroslav, Archangel, Vologda, Novogorod, Kostroma, and Peren.'

At Kargopol he got the information for which he asked, but this concerns us not here. At present we have to do with his journey and with what he saw.

'Village after village passes to the rear. Russ hamlets are so closely modelled on a common type that when you have seen one, you have seen a host; when you have seen two you have seen the whole. Your sample may be either large or small, either log-built or mud-built, either hidden in forest or exposed on steppe; yet in the thousands on thousands to come you will observe no change in the prevailing form. There is a Great Russ hamlet, and a Little Russ hamlet; one with its centre in Moscow, as the capital of Velika Rouss [Great Russia], the second with its centre in Kief, the capital of Malo Rouss [Little Russia.]

'A Great Russ village consists of two lines of cabins parted from each other by a wide and dirty lane. Each homestead stands alone. From ten to a hundred cabins make a village. Built of the same pine logs, notched and bound together, each house is like its fellow, except in size. The elder's hut [Starista] is bigger than the rest; and after the elder's house comes the [Kabac] whiskey-shop. Four squat walls, two tiers in height, and pierced by doors and windows; such is the shell. The floor is mud, the shingle deal. The walls are rough, the crannies stuffed with moss. No paint is used, and the log fronts soon become grimy with rain and smoke. The space between each hut lies open and unfenced, a slough of mud and mire, in which the pigs grunt and wallow, and the wolf-dogs snarl and fight. The lane is planked. One house here and there may have a balcony, a cow-shed, an upper storey. Near the hamlet rises a chapel built of logs, and roofed with plank; but here you find a flush of colour, if not a gleam of gold. The walls of the chapel are sure to be painted white, the roof is sure to be painted green. Some wealthy peasant may have gilt the cross.

'Beyond these dreary cabins lie the still more dreary fields which the people till. Flat, unfenced, and lowly, they have nothing of the poetry of our fields in Sussex and Essex plains; no hedgerow of ferns, no clumps of fruit-trees, and no hints of home. The patches set apart for kitchen stuffs are not like gardens even of their homely kind. They look like workhouse plots of space laid out by yard and rule, in which no living soul had any part. These patches are always mean, and you search in vain for such a dainty as a flower.

. . . .

'The forest melts and melts! We meet a woman driving in a cart alone; a girl darts past us in the mail; anon we come upon a waggon, guarded by troops on foot, containing prisoners, partly chained, in charge of an ancient dame.

'This service of the road is due from village to village; and on a party of travellers coming into a hamlet the elder [Starist] must provide for them the things required —carts, horses, drivers, in accordance with their podorojna; but in many villages the party finds no men, or none except the very young and very old. Husbands are leagues away, fishing in the polar seas, cutting timber in the Kargopol forests, trapping fox and beaver in the Ural mountains, leaving their wives alone for months. These female villages are curious things, in which a man of pleasant manners may find an opportunity of flirting to his heart's content.

'Villages, more villages, yet more villages! We pass a gang of soldiers marching by the side of a peasant's cart, in which lies a prisoner, chained; we spy a wolf in the copse; we meet a pilgrim on his way to Solovetsk; we come upon a gang of boys whose clothes seem to be out at wash; we pass a broken waggon; we start at the howl of some village dogs; and then go winding forward hour by hour, through the silent woods. Some touch of green and poetry charms our eyes in the most desolate scenes. A virgin freshness crisps and shakes the leaves. The air is pure. If nearly all the lines are level, the sky is blue, the sunshine gold. Many of the trees are rich with amber, pink, and brown; and every fragrant breeze makes music in the pines. A peasant and his dog troop past, reminding me of scenes in Kent. A convent here and there peeps out. A patch of forest is on fire, from the burning mass of which a tongue of pale pink flame laps out and up through a pall of purple smoke. A clearing swept by some former fire is all aglow with autumnal flowers. A bright beck dashes through the falling leaves. A comely child, with flaxen curls, and innocent northern eyes, stands bowing in the road with an almost Syrian grace. A woman comes up with a bowl of milk. A group of girls are washing in a stream, under the care of either the virgin mother or some local saint. On every point the folk, if homely, are devotional and polite; brightening their forest brakes with chapel and cross, and making their dreary wood, as it were, a path of light toward heaven

'We dash into a village near a small black lake. . . . This may be such as has been described, for, as has been stated, one general character pervades the whole until you reach the latitude of Kief to the south, and again to the west in which they are different.

'Such is travelling through forest lands in Russia.'

The description is exactly such as I would give if I could; and I wish I could, but I cannot. I felt while I read the narrative as if I could realise it all—the unending road, the weary feeling of being jolted into a jelly; in the blue or green or gilded dome, the village, seen and soon left behind, the travellers overtaken, or met, and passed, the village, and the tearing along of the rough conveyance, with the reins held at arm's length, while the yemshick or driver encourages them with his voice, coaxing, threatening, thrashing, sometimes in the overflowing of his love, and sometimes in fury, and singing the yemschick's song, while the bell on the horses' bow tingles its monotonous accompaniment:

> The troika drives a quiet trot
>   On even road at dead of night;
> The tinkling bell alone doth tell
>   Its near approach, though not in sight.
>   The tinkling bell alone doth tell
>   Its near approach, though not in sight.
>
> The yemschick, roused before the dawn,
>   Feels sadden'd in the chilly night;
> He tries to raise a song in praise
>   Of his village maiden's eyes so bright.
>   He tries to raise a song in praise
>   Of his village maiden's eyes so bright.
>
> "Ye eyes of blue, ye bright blue eyes,
>   Ye've broke a gallant spirit's heart!
> And the oppressors, cruel ever,
>   Have dared two loving souls to part.
>   And the oppressors, cruel ever,
>   Have dared two loving souls to part.
>
> "Fair one! I've left you!" The tears are falling
>   Upon his coat like drops of rain;

>But he dashes off the feeling soft,
>　And the troika drives full speed again.
>　　But he dashes off the feeling soft,
>　　　And the troika drives full speed again.

Or it may be this one :—

>Do you see by the roadside that village ?
>Thither it is that our yemschick is stealing.
>His heart beats quick,
>And he gently sings.
>　His heart beats quick,
>　And he gently sings :
>
>" O thou lovely one that charmed me !
>Now to me the wide world is cold !
>Why ? O why didst thou enchant me ?
>If I was not dear to thy heart ?
>　Why, O why didst thou enchant me ?
>　If I was not dear to thy heart ?
>
>" My darling horses will now languish in grief
>At parting with me, beloved as they are.
>No more will we together tear along
>The verst-post mark'd road.
>　No more will we together tear along,
>　The verst-post mark'd road.
>
>" Not much longer, with wild song
>Not much longer will I cheer the traveller.
>Soon, soon, beneath the sod
>Will lie the body of the young yemshick.
>　Soon, soon beneath the sod
>　Will lie the body of the young yemshick."

And his successor, the driver of the next stage, may sing with laughing irony, which tells that he exults in the assurance that it is not true, or with saddened tone, according as the case may be :—

>Why do you look along the road so eagerly,
>Away from all your cheerful companions ?
>Your heart has begun to beat at the tramping sound !
>Your whole face has become flushed !
>
>Why do you run so hastily
>After the troika which has passed ?
>So, you saw the travelling Count lying on his side in the troika,
>As he gazed on thee as he passed !

No wonder that he looked on thee so intently!
No one can help loving thee.
The yellow robe, and the ornaments in thy black hair
I can liken only to the night.

In the pink hue of thy snowy face
I see the dawn coming forth;
From under thy arched eyebrows
There looks forth the roguish eye!

One look of the dark, broad-shouldered gipsy,
So like the spark of fire setting on fire the forest,
Would cause the old man to ruin himself in presents,
But in the heart of the young man awakens only love!

Yes, you will have lovers enough and to spare,
Thy life will be both full and free,
To thy share will fall abundance,
To the pawky slut comes not the poor *mujik* (peasant).

Such is the yemschick, one of the most interesting, if not the only interesting, object of study on such a journey. He may be clad in a sheepskin shoube, or in hodden grey, or, like some village Lothario, in velvet coat, loose red drawers, and shining boots, with one or two, or it may be half a dozen peacock feathers wound round his hat—the indication of his being a Government driver. Such are the yemschick, his sentiments and his songs, these being generally of love, and always in a minor key, wound up and closing with a long drawn out fugue, dolorous, plaintive, rising and swelling and dying away with the cadence of an Æolian harp; and then there follows generally a word or two of endearment addressed to his horses—one of them addressed as his *golubchik* or turtle dove, and another as his *doushinka* or sweet little soul; but sometimes words of scolding are used, such as I would not willingly repeat, save to tell that during the time of the Crimean war, and for some time thereafter, they had no more spiteful names to call them than Palmerston and Aberdeen!

CHAPTER VIII.

LAPLAND, AND LAND OF THE SAMOIDES.

From about the latitude of Archangel, but on the western coast of the White Sea, and extending thence to the frontier of Finland, is Russian Lapland. It also is wooded, but the country inhabited by the Lapps extends through Finland, Sweden, and Norway, to the Atlantic, and the timber trade is much more extensively developed in the Scandinavian portion of Lapland than in that which lies further to the west. The Tornea flowing into the head of the Gulf of Bothnia, and the western boundary of Finland, is considered the medium line.

Lapland has been divided by Wahlenberg into five zones, concentric with the Gulf of Bothnia, and differing from each other in climate and productions. The first, extending obliquely round the Gulf of Bothnia, from N. lat. 64° to nearly 69°, and forming a zone generally 80 miles in breadth, is covered with forests of spruce and Scots fir, and is called Woody Lapland. The second, higher and colder than the first, extending from latitude 65° to nearly 70°, and generally only six or eight miles in breadth, contains the Scots fir, and is denominated Sub-woody Lapland. The third, of a higher elevation than either of the others, ranges, like the latter, from 65° to 70° N. lat., and is generally about twelve miles in breath, except to the north-east of Enonteki, where it descends to about 40°, produces the birch, but none of the conifers; it is called the Sub-Alpine region. The fourth, immediately behind the third, and nearly of the same breadth, stands still higher, and produces only the *Salix glauca,* a species of willow peculiar to very cold climates; it is called the Lower Alpine region. The fifth, the Higher Alpine region, lies beyond this. Much of it is covered with perpetual snow; it produces no trees, and scarcely any vegetation whatever, except a few hardy plants where the snow has disappeared.

Mr A. G. Guillemard, writing in the *Journal of Forestry* of September 1882, of Forest Rambles in Swedish Lapland, tells :—

'An almost unbroken solitude of vast forests and wide-spreading moorland, lonely lakes, and rushing rivers, with lofty ranges of noble snow-mountains in the far interior, dividing the watershed between the Gulf of Bothnia and the Atlantic Ocean; a "wild north land" in which the solitary traveller from countries of so-called civilisation is regarded with eyes of wonderment, and to which the modern tourist never comes; the home of the elk and the bear, the ptarmigan and capercailzie—such is the land of the Lapp. A glance at the map of the country will suffice to convince the intending traveller that this far-distant land is eminently a land of waters, rivers innumerable flowing through long strings of lakes forming its main characteristic. But the four days' voyage from Stockholm to Lulea at the head of the Gulf of Bothnia will speedily lead him to conclude that it must be a land of trees as well, for almost every ship he meets is a timber ship, every port at which his steamer touches is crowded with rafts of pine-trunks, and noisy with saw-mills, and the entire coastline is densely covered with pine forest. His deduction will prove to be a correct one, though the sights and sounds from which he draws it are Swedish, the boundary of Lapland being far up country. The cream of the forest and mountain and lake scenery, and perhaps of the sport as well, is to be found in the vicinity of the Stora Lule river and its tributaries, a truly noble stream, which for many miles above its mouth averages more than a mile in width. Small steamers navigate the Stora Lule for a distance of a hundred miles, and at a point just ninety miles from Lulea, a lane cut through the forests down to the river banks marks the boundary line of Lapland.

'Throughout the passage up stream the fact that the main product of the interior of Lapland is timber is plainly evidenced by the vast number of pine logs floating down with the current. Felled far away in the recesses of the forest, and roughly trimmed of their branches, the trunks are hauled to the nearest stream during the early spring, so that when the summer suns have melted the snows on the mountains, and unbound the icy fetters on lake and river, the fruits of the winter's forestry are borne seaward on the rushing flood. Throughout the summer and autumn, and up to mid-November, by which date the Lule is generally frozen up, the endless procession of logs continues. Although the floats of the paddles of steamers are guarded by chains, the careful helmsman generally gives the wheel a spoke or two to avoid trunks immediately ahead, exercising similar discretion to that evinced by the quartermaster of a North Sea steamer in steering clear of the far-spreading nets of the fishing fleet on the Dogger Bank. These trunks are all marked, so that they can be properly identified by gangs of out-lookers at the mouth of the river, who are always on the alert to secure them, and consign them to the respective saw-mill owners or shippers for whom they have been despatched. All along the course of the river, too, men are employed by the Government to maintain the traffic unimpeded, to clear "jams" of logs in the rapids, and to set afloat such as may have drifted ashore and been left high and dry as the river decreased in volume.

'At Storbacken, a hundred miles up-stream, steamer navigation comes to an end, the rapids of Porsi-forssen presenting an insurmountable obstacle. Hence a drive of thirty-two miles, following the course of the Lilla Lule through undulating forest country, diverisfied here and there by small clearings, where good crops of potatoes, barley, and oats please the eye of the farmer, takes the traveller to Jokkmokk, the nearest of Lapp villages to civilisation. The scanty population of this part of the country may be judged from the fact that in September last I drove the entire distance without meeting a single vehicle on the road. Most of the farms are to the right of the road along the river bank; to the left one might strike into a pathless forest and wander aimlessly for days together with but a slight chance of lighting upon a house or even a sign of a human being. At one part of the road I drove for three miles through a wilderness of gaunt blackened trunks of pines, across which a forest fire had swept some three years previously. In Tasmania, New South Wales, and California, I had passed through similar scenes of desolation, and the surroundings are always eerie and depressing, but, like most gloomy things in this world, they have redeeming features. The wondrous after-glow of these high latitudes showed up in strong relief the naked ruined trunks, and the havoc that had been wrought was vividly portrayed; but at the same time the growth of the new forest, the young pines and birches, the largest of them already from five to eight feet in height, and the vigorous undergrowth, afforded ample evidence of the recuperative properties of forest soil, only requiring time to develop themselves in order to replace the old forest that had been swept away.'

We should err if we were to conclude from this that Lapland, even in its wooded parts, is one continuous forest such as M. Judræ traversed in the Government of Olonetz, and Mr Hepworth Dixon in the Governments of Archangel and Vologda. The soil of Lapland is generally sterile. The greater part is covered with rocks, or moss, or gravelly plains, or a kind of turf composed of mosses destroyed by the frost, and impregnated with stagnant water; the variety of vegetation is more striking than its abundance. Wahlenberg's edition of the *Flora Lapponica* gives descriptions of 1087 species of plants; of which 496 are phanerogamous and 591 are cryptogams. Of trees there are 26 kinds, consisting af the Scots fir, spruce fir, birch, alder, mountain ash, birdcherry, and nineteen species of willow.

Mr Guillemard and a fellow traveller fixed on Jokkmokk as their headquarters, and thence made daily excursions into the forest, visiting the neighbouring lake and river in quest of trout and grayling.

'Between us,' says he, 'we had travelled in all the four quarters of the globe, and, with eyes for the picturesque, were of opinion that a fairer landscape than that viewed from the road a mile or so south-east of this little Lapp village, could not be desired even by a Ruskin. A low wooden bridge spans a brawling torrent, eddying in rapids through peaty brown pools, and mingling its waters with those of the Lilla Lule, which here expands into a broad reach, a mile across, studded with two or three islets fringed with reeds, and crowned with crests of pine and yellow birch. Its waters are of a deep glowing blue, reflected from the brilliant sky overhead, and contrasting vividly with the flame-colour of the birch leaves and the sombre dark green of the pines. To the right is a tiny clearing, on which a crop of oats is fast ripening under the hot suns of autumn, but this alone gives sign of the handiwork of man. From the opposite bank of the river rises a steep slope of forest, and each headland and curve of bay is thickly covered with pine and fir, and birch and alder, which, touched by the sharp frosts of the September nights, already warn us of the waning of the year. In the pools beneath the bridge, on the placid surface of which the sun strikes hotly, we can plainly see a hundred tiny fish holding high revel amongst the boulders which form the bed of this Devonshire-like trout-stream; but away in the great river before us, running in swirling eddies round the headlands, and wildly tossing tall rainbow-tinted jets of spray high into the sunny air amongst the rapids between the islands, the speckled trout and grayling have their home, and the lordly salmon cruises regardless of fly or minnow. Whether seen under the full blaze of noon-tide sunshine, or the wondrous brilliance of the Northern Lights, this panorama of river and forest scenery never failed to charm us.

'I have paid many visits to Scandinavia, but cannot remember any forest ramble so replete with all the numerous beauties of nature and scenery as that taken by Mr D. and myself on our expedition to the Falls of Njommelsaska in September last. These superb falls are distant thirty-five miles from Jokkmokk, and lying as they do far to the north of the route from Lulea to Quikkjokk, and entailing on the traveller a seventy mile walk through a wild forest country, are but seldom visited. In the entire Continent of Europe there is no cataract of equal volume and grandeur; and yet so remote is it from the haunts of man, that we learned at Jokkmokk that no one had visited Njommelsaska for two years prior to our arrival. The lion of European waterfalls cannot complain that the solitudes in which he lies hidden are ever disturbed by the incursion of " specially-conducted parties."

'Having despatched some hours previously a sturdy young Swede with our ulsters and hammocks, to give warning of our advent at the half-way house, where we were to obtain quarters for the night, we started at midday, knapsacks on backs, for Vajkijaur, the first of the long string of superb lakes which stretch away westward for sixty miles to Quikkjokk. Whilst rowing across its blue waters, the grand snowy peaks of the Kabbla mountains charm the traveller with their picturesque outline, and indicate the distant spot where Suoloitjelma, king of northern ranges, lifts it proud peak above untrodden snows. We land at a little hamlet at the foot of a southern slope covered with small fields of grain and potatoes, and without loss of time strike into the forest to the northward. The trail leads through a dense growth of Scots fir for several miles, the trees being of small girth, and in but few instances exceeding eighty feet in height. So thick is the forest that the branches overhead form almost a complete ceiling, and the ground is so plentifully strewn with the fallen needles that the path is barely traceable. A great stillness reigns, the only sound noticeable being the low murmur of the breeze through the foliage far above us. The trail then gradually rises, the fir forest is left behind, and we pass through scant birch and alder shrubbery, varied occasionally by long stretches of marsh, hemmed in by tracts of reeds from six to eight feet in height, from out of which the unwonted sound of our voices scares a small string of duck. These marshy tracts are common all over Scandinavia, which is merely a vast substratum of rock covered with a shallow soil. Huge tussocks of coarse grass, rushes, and diminutive shrubs afford precarious foothold: these are covered in many places with the foliage of water-plants, whilst the boulders, which crop up in every direction, are overgrown with delicate mosses and lichens of a hundred tints of green, gray, and brown.

'We then re-enter the forest and dive into a lovely valley, the floor of which is emerald green, with a thick carpet of grass, from out of which springs here and there a tiny specimen of the oak fern, a rarity in these high latitudes. Here we come across some grand spruce firs, ranging to a hundred feet in height and of perfect symmetry, and the sylvan beauty of the scene is enhanced by a tiny brook coursing away over a boulder-strewn bed between banks of tall grasses and the fleecy white seed-spikes of the cotton plant. A steep slope of some five hundred feet of ascent closes in the dell to the north, and as we are toiling up the rugged path a couple of capercailzie soar slowly upwards into the blue sky and are lost to sight over the tops of the firs. The summit gained, a magnificent prospect is before us, the broad blue expanses of Lakes Vajkijaur, Purkijaur, and Randijaur lying as it were at our feet, with the rapids of the Lilla Lule plainly visible amongst the dark forest round Jokkmokk, and the Kabbla snows glistening white on the far horizon. Around us rise the ruddy trunks of pine and fir, the sombre foliage of which is abundantly diversified by the flaming yellow and scarlet of the birches and aspens, and the brilliant crimson of the shrubs of rowan, whilst the mosses, lichens, and tiny forest-plants form a perfect mosaic of rich colouring under our feet. Hence a gradual descent through thick birch shrubbery leads us to Anajaur, a lake some four miles round, embosomed in forest, and famous for the size of its pike, which, however, are but seldom disturbed by the drag-nets of the Ligga farmer. Having crossed this in a crazy boat, an hour's walk through spruce forest brings us to the tiny clearing we seek.

'Ligga is a small farm inhabited by a hardy Swede, who has cleared away some twenty acres of forest on a plateau about five hundred feet above the Stora Lule river, which, having left at its junction with the Lilla Lule at Posi-forssen, we now strike some forty miles nearer its source. The farm-house consists of the customary weather-board building of two rooms, and is prettily placed close to the bank of a mountain stream rushing down in cascades to join the big river at the foot of the mountain slope. One of these rooms is allotted to us, and in this we sling our hammocks, and, in spite of the cold, for chinks are numerous in the walls, and it is freezing hard outside, pass a fairly comfortable night. The ground is sparkling with hoar-frost when we turn out next morning, and, greatly to the amusement of a couple of Lapps, the labourers at the farm, proceed to make our ablutions in a big pool of the torrent. A hearty breakfast of Chicago beef, biscuits, and coffee—the latter always excellent in Scandinavia—and at nine we start for the falls. This part of the country is almost uninhabited—we are deep in the recesses of the forest primæval, with a plethora of the beauties of nature round us, rendered doubly beautiful by the aspect of all-pervading repose and solitude. Save at Ligga we see not a single human being in the course of our seventy mile walk. We are on higher ground this second day, and some of the firs and spruces are of noble dimensions, but the under-growth is scantier, owing to the more exposed nature of the country and the shallowness of the soil. Every-where huge boulders lie in picturesque confusion, and the gnarled roots of the trees twist and curl here and there seeking in vain for sufficient depth of soil to cover them.

Under these conditions, and remembering that we are far within the Arctic Circle, and that the snowfall and the gales of winter must severely try the stability of the trees, the number of fallen giants is not to be wondered at. But though the floor of the forest wears a somewhat more desolate aspect than that between Vajkijaur and Ligga, the general tone of colour being gray in place of the green of the grasses and mosses of the lower country, the dark glossy leaves and crimson clusters of the *molte-baere*, the bright crimson of the tiny shrubs of rowan and the deep red-brown of the lichens spreading over the larger masses of rock diversify the colouring of the landscape pleasantly. In places we come across a small stream tinkling musically over the boulders between banks fringed with aspen and birch thickets, and in secluded spots find the Arctic strawberry flourishing, and bearing a rich crop of delicious aromatic fruit. A walk of ten miles brings us to the Stora Lule again, here a river of some two hundred yards across, its deep blue waters flecked with huge masses of creamy foam, the product of the mighty fall, the thunder of which, echoing from out the deep canyon in which it is secluded, now falls plainly on our ears. We find a boat hauled up in a quiet nook among the rocks, and paddle across to the left bank of the river, from which the Ananas Mountains rise in a steep slope of birch forest, now ablaze with the golden splendour of autumnal tints. Hence to the foot of the falls is a distance of some two miles, at first a walk over a smooth lawn of short sweet grass, on which we find the poles, erected in circular shape, of a Lapp encampment, and finally a rough scramble over huge masses of rock under the lee of which thickets of Arctic raspberries afford us a rich feast, the delights of which, however, are a trifle marred by the clouds of mosquitoes and midges which as yet have survived the early frosts.

'As we gaze down into the deep canyon, and trace the course of the river from the broad blue reach above the falls, over the three great plunges and down the series of rapids by which they are connected, surveying the whole six hundred yards of cataract in its various phases of sheer fall, boiling cauldron, wave and eddy and rapid, with the spray flying over us in clouds hurled up forty feet into the rainbow-sprinkled air from where the huge waves are breaking in thunder on the ink-black cliffs below us, we find no difficulty in realising that the grandest fall in Europe is before us. The river is about 150 yards across above the crest of the first fall, when at high flood, after the melting of the snows, and some thirty feet in depth; but now, in early autumn, its volume is much reduced, and huge black rocks divide it into three separate streams at the lip. Some idea of the stupendous rush of waters may be judged from the fact that this superb river falls 250 feet in the course of six hundred yards. Immediately below the first fall of forty feet is an inaccessible islet, covered with pines and birches, the leaves of which quiver with the concussion of the falling waters, just as those of the shrubs on Luna Island, hanging on the brink of Niagara, are ever tremulous.

'So indescribably grand is Njommelsaska, that it is with the utmost reluctance that we turn our backs upon its glories, and wend our way back to Ligga through the silent forest, whence a third long day's walk brings us to Jokkmokk.'

This, however, though within woody Lapland, lies far to the west of the line which divides Scandinavian Lapland from Russian Lapland.

The whole of woody Lapland is so level that scarcely one of the mountains rises higher than 213 feet above the neighbouring plains, and in none of the first three zones is the height above the level of the sea considerable, but a few high mountains there are. And the Lapland Alps have an altitude such that no part is less than 2000 feet above the level of the sea.

The fifth region, the higher Alpine region, extends along the north side of Lapland, varying in breadth as it may happen to be indented by the sea. In the southern Alpine region there are mountains and glaciers 4000, 5000, and 6000 feet above the level of the sea. Of the maritime Alps, which occupy the west and northern part of Lapland, and which has glaciers immediately over the sea, the highest are the Alps of Lyngen, which rise to an elevation of 4264 feet. The rest of the coast of Lapland is very rocky; but, excepting the promontory of Kunnen, it scarcely contains any high mountains. The promontories of eastern Finmark do not exceed an elevation of 2132 feet above the level of the sea; and those on its north coast are only 1279 feet in height; and a long stretch of comparatively level land is presented by the coast of the Arctic Ocean in Russian Lapland. But it is begirt by a mountain range, on the south of the coast of the White Sea. Here it is that the forests commence. Much of the land to the north of the Arctic Circle, like the land in the same latitude to the east of the White Sea, resemble the barrens of North America, and the *Tundras* of Siberia, sterile marshy wastes.

Of those lands washed by the White Sea the following account is given by Hepworth Dixon, who wielded a graphic pen perhaps somewhat freely, but if so all the more effectively, in bringing before his readers the scenes he describes. He is describing his approach to Russia, to which he went by Archangel, in preference to taking any of the more generally adopted routes in the south.

'Rounding the North Cape, a weird and hoary mass of rock projecting far into the Arctic foam, we drive in a south-east course, lashed by the wind, and beaten by hail and rain, for two long days, during which the sun never sets and never rises, and in which if there is dawn at the hour of midnight, there is also dusk at the time of noon.

'Leaving the picturesque lines of fiord and alp behind, we run along a dim, unbroken coast, not often to be seen through the pall of mist, until, at the end of some fifty hours, we feel, as it were, the land in our front; a stretch of low-lying shore in the vague and far-off distance, trending away toward the south, like the trail of an evening cloud. We bend in a southern course between Holy Point (Sviatoi Noss, called in our charts, in rough salt slang, Sweet Nose) and Kanin Cape, towards the Corridor; a strait of some thirty miles wide, leading from the Polar Ocean into that vast irregular dent in the northern shore of Great Russia, known as the Frozen Sea.

'The land now lying on our right, as we run through the Corridor, is that of the Lapps; a country of barren downs and deep black lakes; over which a few trappers and fishermen roam; subjects of the Tsar, and followers of the orthodox rite; but speaking a language of their own, not understood in the Winter Palace, and following a custom of their fathers, not yet recognised in St. Isaac's Church. Lapland is a tangle of rocks and pools; the rocks very big and broken, the pools very deep and black; with here and there a valley winding through them, on the slopes of which grows a little reindeer moss. Now and then you come upon a patch of birch and pine. No grain will grow in these Arctic zones, and the food of the natives is game and fish. Ryebread, their only luxury, must be fetched in boats from the towns of Onega and Archangel, standing on the shores of the Frozen Sea, and fed from the warmer provinces in the south. These Lapps are still nomadic, cowering in the winter months in shanties; sprawling through the summer months in tents. Their shanty is a log pyramid, thatched with moss to keep out wind and sleet; their tent is of the Comanche type; a roll of reindeer skins drawn slackly round a pole, and open at the top to let out the smoke.

'A Lapp removes his dwelling from place to place, as the seasons come and go; now herding game on the hillsides, now whipping the rivers and creeks for fish; in the warm months roving inland in search of moss and grass; in the frozen months drawing nearer to the shore in search of seal and cod. The men are equally expert with the bow, their ancient weapon of defence, and with the birding piece they warn off settlers in their midst. The women, looking anything but lovely in their sealskin tights and reindeer smocks, are infamous for magic and second sight. In every district of the north a female Lapp is feared as a witch—an enchantress, who keeps a devil at her side, bound by the powers of darkness to obey her will. She can see into the coming day. She can bring a man ill luck. She can throw herself out into space, and work upon ships that are sailing past her on the sea. Far out in the polar brine, where her countrymen fish for cod, stands a lump of rock, which the crews regard as a woman and her child.

'Such phantasies are common in these Arctic seas where the waves wash in and out through the cliffs, and rend and carve them into wondrous shapes. A rock on the North Cape is the Friar; a group of islets near that cape is known as the Mother and her Daughters. Seen through the veil of Polar mist, a block of stone may take a mysterious form; and that lump of rock in the Polar waste, which the cod fishers say is like a woman with her child, has long been known to them as the Golden Hag. She is rarely seen; for the clouds in summer, and the snows in winter, hide her charms from the fisherman's eyes; but when she deigns to show her face in the clear bright sun, her children hail her with a song of joy, for on seeing her face they know that their voyage will be blessed by a plentiful harvest of skins and fish.

'Woe to the mariner tossed upon their coast!

'The land on our left is the Kanin Peninsula; part of that region of heath and sand over which the Samoyed roams; a desert of ice and snow still wilder than the countries hunted by the Lapp. A land without a village, without a road, without a field, without a name; for the Russians who own it have no name for it save that of the Samoyed's land. This province of the great empire wends away north and east from the walls of Archangel, and the waters of the Kanin Cape to the summits of the Ural chain, and the iron gates of the Kara Sea. In her clefts and ridges snow never melts; and her shorelands, stretching toward the sunrise upwards of two thousand miles, are bound in icy chains for eight months in the twelve.

'In June, when the winter goes away suddenly, the slopes of a few favoured valleys grow green with reindeer moss; slight spects of verdure in a landscape which is even then dark with rock and grey with rienne. On this green moss the reindeer feed, and on these camels of the Polar Zone, the wild men of the country live.'

CHAPTER IX.

NOVA ZEMBLA AND LANDS BEYOND.

The preceding notice of the Land of the Samoides, to which we have been led by taking notice of the northern coast lands of Lapland, has led us out from the forests, if not beyond the forest-lands, of Northern Russia; and having ventured so far, surely we may venture across the narrow strait beyond, and take a glance at Nova Zembla, or Novaia Zemblia—the New Land—before we betake ourselves to the study of the forest economy of the land whither we have found our way. This projects from the most northern point of Russia in Europe, near to its eastern boundary, from which it is separated by the Waygatz Shoals. With an exception, which will afterwards be referred to, it may be said to be uninhabited; but it is visited by fishermen and hunters, who are sent out by the merchants of Archangel and Mezen to obtain whales and walrusses. It is generally spoken of as one island, but being traversed by a narrow crooked passage from west to east, there are two large islands, with some lesser ones on the coast. Coal and asphaltum have been found in the interior, and there exists a salt lake there.

A writer in *Blackwood's Magazine*, in the issue for September 1883, who appears to be keenly alive to the pleasure experienced in the chase, and who came home with the crew of the ill-fated 'Eira,' supplies some interesting details in regard to this land, amongst others these :—

'Being far out of the way of all our merchant routes, and only approachable during the summer over the, even then, ice-encumbered sea, Nova Zembla will probably long remain one of the last refuges of the reindeer; while its ice-choked fiords and frozen seas will still be haunted by the white whale, the seal, the walrus, and the polar bear.

'Frequented until of late only by some dozen schooners, who visit its shores every year chiefly for white whale and salmon, and by a few roaming Samoides from the mainland, these Arctic shores have hitherto afforded an undisturbed asylum during the winter to game of all kinds, marine or terrestrial, which abounds there. Recently, however, the Russian Government have seen fit to plant a colony of Samoides, and these skilful hunters harry the game throughout the year with great vigour. Beyond visits from European sportsmen or explorers, so rare that they might almost be counted on the fingers, no other human intruders ever invade these wild regions. . . . .

'Till the present century the contour of the two large islands which form what is now known as Nova Zembla was very differently represented upon the various manuscript charts in existence, these having been compiled from the observations of Dutch, Norwegian, and Russian navigators. Barents led off in 1598 with a chart representing the west coast and that part of the north-east coast which he had visited; this, though terribly out in longititude, was very good as to latitude, and since the days of this old explorer, his maps, with many additions and a few corrections, have been generally adhered to: some representing the north coast as taking an abrupt turn to the east, and thus continuing *ad infinitum*, the authors veiling their perplexity by drawing a meridian line down the chart and thereby cutting it short, leaving the rest to the imagination of the beholder.

'For our present knowledge of the shape and dimensions of the islands we are chiefly indebted to the Russian coast survey made during the early part of the present century, and continued by subsequent explorers, which is generally considered to be pretty accurate as far north as Admiralty Peninsula, the most prominent headland on the west coast on the north island. . . . . Cape Nassau, the point between Admiralty Peninsula and Cape Mauritius, the north point has traditionally acquired an evil reputation amongst the walrus hunters, as being a sort of bewitched headland, to round which means to say farewell to the world; for it was believed that vessels were mysteriously drifted thence into the Arctic Ocean, beset by ice, and never heard of again. That there is some foundation for this tradition, is proved by the fate of the Austrian Polar expedition of Weyprecht and Payer in the steamer Tegethoff, which was beset near this Cape in 1872, and never got free again, being drifted about the Arctic Ocean for two years, during which the expedition involuntarily discovered Franz-Josef land, and only at last got free by abandoning their ship, and undertaking a most perilous and laborious journey over the ice with their boats, which lasted three months, when they had the good fortune to reach the shores of Nova Zembla, and to encounter a Russian schooner which was just leaving for home.

'Lying between the parallels of 77° 35′ N. and 70° 40′ N., the two main islands, with a curved outline, cover a space,' says our author, 'of about 450 English miles, while their average breadth may be taken as 60 miles. The two islands are divided by a strait called the Matotchkin Sharr, which also well marks a central position in the physical configuration of the country; for it is in this locality that the highest mountains and wildest scenery are to be found, the land thence sinking to lower levels both to the northward and southward. Matotchkin Sharr may likewise be said to be a central position as to the distribution of the various objects of sport; for it is on the slopes of the snow and glacier clad mountains of this part of the country that reindeer are most plentiful, whilst wildfowl of all kinds prefer the south island. Bears, walrusses, and seals, on the other hand, may be looked for with greater confidence on the shores of the north island, and more particularly on the eastern and northern parts of it.'

Mention is made of Barents, and of the Austrian expedition under Payer.

'The merchants of Amsterdam,' writes the author of *The Arctic World: its Plants, Animals, and Natural Phenomena,** 'having fitted out a ship—the *Mercurius*, of one hundred tons—to attempt a passage round the northern end of Novaia Zemlaia, the command was given to William Barents; who accordingly sailed from the Texel on the 4th of June 1594.

'He sighted Novaia Zemlaia, in lat. 73° 25′ N., on the 4th July, sailed along its grim, gaunt coast, doubled Cape Nassau on the 10th, and struck the edge of the northern ice on the 13th. For several days he skirted this formidable barrier, vainly seeking for an opening; and in quest of a channel into the further sea he sailed perseveringly from Cape Nassau to the Orange Islands. He went over no fewer than seventeen hundred miles of ground in his assiduous search, and put his ship about one-and-eighty times. He discovered also the long line of coast between the two points we have named, laying it down with an exactness which has been acknowledged by later explorers. His men wearying of labour which seemed to yield no positive results, Barents was under the necessity of returning home.

'In 1596 the Amsterdammers fitted out another expedition, consisting of two strongly-built ships, under Jacob van Heemskerch and Jan Cornelizoon Rijp, with Barents as pilot, though really in command.

'In this voyage the adventurers kept away from the land in order to avoid the pack-ice, and sailing to the westward, discovered Bear Island on the 9th of June. Then they steered to the northward, and hove in sight of Spitzbergen exactly ten days later. They supposed, however, that it was only a part of Greenland, and were led to bear away to the north-west—a course which was speedily arrested by the eternal icy barrier. Barents then coasted along the western side of Spitzbergen; and the north-western headland being frequented by an immense number of birds, he called it Vogelsang.

---

* T. Nelson & Sons, London, Edinburgh, and New York.

'On the 1st of July he again made Bear Island, and here he and Rijp agreed to separate. Of the latter we know only that he was unsuccessful in an attempt to find an opening in the ice on the east of Greenland, and that he returned to Holland in the same year. Of the former the narrative is painfully full and interesting.

'Quitting Bear Island, he reached Novaia Zemlaia on the 17th of July, sighting the coast in lat. 74° 40′ N. Keeping along it with characteristic perseverance until the 7th of August, he passed Cape Comfort; but only to find himself once more face to face with the dreary spectacle of the far-reaching Polar ice. It so hemmed and fenced him in on every side that he was unable to extricate his vessel from it; and being driven into a bay, which he named Ice Haven, "there they were forced, in great cold, poverty, misery, and griefe, to stay all the winter." For the heavy pack-ice drifting into the bay closed it up, and closed around the ship until she was held fast as in iron bonds.

'Barents and his sixteen followers now prepared to encounter with a good heart the trials of the long Arctic winter night. They displayed, in truth, a courage, a patience, and a good fellowship which were heroic. Finding a large supply of drift-wood, they constructed, with the help of planks from the poop and forecastle of the vessel, a sufficiently commodious house, into which they removed all their stores and provisions They fixed a chimney in the centre of the roof; a Dutch clock was set up, and duly struck the weary hours; the sleeping-berths were ranged along the walls; a wine-cask was converted into a bath. All these ingenious devices, however, availed but little against the terrible feeling of depression which is induced by the continuance for so many weeks of a blank and cheerless darkness.

'The sun disappeared on the 4th of November, and the cold thereafter increased until it was almost intolerable. Their wine and beer were frozen, and lost all their strength. By means of great fires, by applying heated stones to their feet, and by wrapping themselves up in double fox-skin coats, they barely contrived to keep off the deadly cold. In searching for drift-wood they endured the sharpest pain, and often braved imminent danger. To add to their troubles, they had much ado to defend themselves against the bears, which made frequent assaults on their hut. However, they contrived to slaughter some of the audacious animals, and their fat provided them with oil for their lamps. When the sun disappeared the bears departed, and then the white foxes came in great numbers. They were much more welcome visitors, and being caught in traps, set in the vicinity of the house, supplied the ice-bound voyagers with food and clothing.

'When the 19th of December arrived, they found some comfort in the reflection that half of the dreary season of darkness had passed away, and that they could now count every day as bringing them nearer to the joyful spring. They suffered much, but endured their sufferings bravely; and celebrated Twelfth Night with a little sack, two pounds of meat, and some merry games. The gunner drew the prize, and became King of Novaia Zemlaia, "which is at least two hundred miles long, and lyeth between two seas."

'On the 27th of January every heart rejoiced, for the glowing disc of the sun reappeared above the horizon. But it brought with it their old enemies the bears, against whom they found it necessary to exercise the greatest vigilance.

'On the 22nd of February they again saw "much open water in the sea, which in long time they had not seene." During the whole month violent storms broke out, and the snow fell in enormous quantities.

'On the 12th of March a gale from the north-east brought back the ice, and the open water disappeared; the ice driving in with much fury and a mighty noise, the pieces crashing against each other, "fearful to hear." Up to the 8th of May the ice was everywhere, and their sad eyes could look forth on no pleasant or hopeful scene. Then it began to break up, and the gaunt, weary explorers prepared to tempt the sea once more. They set to work to repair their two boats, for their ship was so crippled and strained by the ice that she was injured beyond their ability to repair.

'On the 14th of June they quitted the place of their long captivity; Barents, before they set out, drawing up in writing a list of their names, with a brief record of their experiences, and depositing it in the wooden hut. He himself was so reduced with sickness, want, and anxiety that he was unable to stand, and had to be carried into the boat. On the 16th, the captain, hailing from the other boat, inquired how the pilot fared. "Quite well, mate," Barents replied; "I still hope to mend before we get to Wardhouse,"—Wardhouse being an island on the coast of Lapland. But he died on the 19th (or, as some authorities say, on the 20th), to the great grief of his comrades, who appreciated his manly character, and placed great reliance on his experience and skill.

'The adventurers met with many difficulties from the ice,—sometimes being carried out far from the ice-belt, and at others being compelled to haul the boats for long distances over the rough surface of the floes to reach open water. It has been well observed that there are many instances on record of long ocean-voyages performed in open boats, but that, perhaps, not one is of so extraordinary a character as that which we are describing,—when two small and crazy craft ventured to cross the seas for eleven hundred miles, continually endangered by huge floating ice-masses, threatened by bears, and exposed for forty days to the combined trials of sickness, famine, cold, and fatigue.

'At length they arrived at Kola, in Lapland, towards the end of August; and, strangely enough, were taken on board a Dutch vessel commanded by the very Cornelizoon Rijp, who had commanded the sister discovery ship in the previous year. They reached the Maas in safety in October 1597.

'No voyager appears to have sailed in the track of

Barents, or, at all events, to have visited the place where he wintered, until 1871. No one but he had rounded the north-east point of bleak Novaia Zemlaia. In 1869, however, and on the 16th of May, Captain Carlsen, a Norwegian of much experience in the North Sea trade, sailed from Hammerfest in a sloop of sixty tons, called the *Solid*. On the 7th of September he reached Ice Haven, and on the 9th discovered a rude wooden house standing at the head of the bay. Its dimensions were 32 feet by 20, and it was constructed of planks measuring from 14 to 16 inches in breadth, and 1½ inches thick. These, it was evident, had belonged to a ship, and amongst them were several oak beams. Heaps of bones of seal, bear, reindeer, and walrus, as well as several large puncheons, were collected round the hut. It was the winter prison of Barents and his companions, and had never been entered by human foot since they had abandoned it. The cooking-pans stood over the fireplace, the old clock hung against the wall; there were the books, and implements, and tools, and weapons, which had been of so much service two hundred and seventy-eight years before. It was an Arctic reproduction of the legend of the hundred years' sleep of the fairy princess.

'Captain Carlsen gives the following list of articles found in the lone hut on the shore of Novaia Zemlaia:—Iron frame over the fireplace, with shifting bar; two ship cooking-pans of copper, found standing on the iron frame, with the remains of a copper scoop; copper bands, probably at one time fastened round pails; bar of iron; iron crowbar; one long and two small gun-barrels; two bores or augers, each three feet in length; chisel, padlock, caulking iron, three gouges, and six files; plate of zinc; earthenware jar; tankard, with zinc lid; lower half of another tankard; six fragments of pepper pots; tin meat-strainer; pair of boots; sword; fragments of old engravings, with Latin couplets underneath them; three Dutch books; a small piece of metal; nineteen cartridge cases, some still full of powder; iron chest, with lid, and intricate lockwork; fragments of metal handle of same; grindstone; an eight-pound iron weight; small cannon-ball; gun-lock, with hammer and flint; clock, bell of clock, and striker; rasp; small auger; small narrow strips of copper band; two salt and pepper pots, about eight inches high; two pairs of compasses; fragment of iron-handled knife; three spoons; borer; hone; one wooden, and one bronze tap; two wooden stoppers for gun muzzles; two spear or ice-pole heads; four navigation instruments; a flute; lock and key; another lock; sledge-hammer head; clock weight; twenty-six pewter candlesticks and fragments, six in a complete state of preservation; pitcher of Etruscan shape, beautifully engraved; upper half of another pitcher; wooden trencher, coloured red; clock alarum; three scales; four medallions, circular, about eight inches in diameter, three of them mounted in oak frames; a string of buttons; hilt of sword, and a foot of its blade; halberd head; and two carved pieces of wood, one with the haft of a knife in it.

'On the 14th of September Captain Carlsen sailed from the Ice Haven, and kept along the east coast of Novaia Zemlaia, encountering bad weather and contrary winds, but succeeding in his chief object, the circumnavigation of the island, which he accomplished on the 6th of October. He returned to Hammerfest early in November.

'Our chronological summary now brings us to the Austrian Polar expedition of 1872. The command was intrusted to Lieutenant Payer, an accomplished seaman who had served under Captain Koldewey; Carlsen was engaged as pilot; and the steamer *Tegethoff* was carefully and abundantly equipped for the voyage. The plan laid down by Lieutenant Payer was well conceived; namely, to round the north-eastern point of Novaia Zemlaia, and sail eastward until he made the extreme north of Siberia, where he proposed to winter. In the spring, travelling parties would be sent out on exploring journeys; and the voyage in summer would be continued as far as Behring's Strait.

'The *Tegethoff* steamed out of Tromsö Harbour on the 13th of July; first fell in with the ice on the 25th, in lat. 74° 15′ N.; and on the 29th sighted the coast of Novaia Zemlaia. Here she was caught in the pack, but steam being got up, repeated charges were made at the enemy, and she was carried bravely into an open water-way, about twenty miles wide, to the north of the Matochkia Strait. On the 12th of August she was joined by the *Isbyörn* yacht with Count Wilczck and some friends on board. The two vessels anchored close to the shore in lat. 76° 30′ N., and on the 18th celebrated the Emperor of Austria's birthday. Daily excursions were made by sledge parties to the adjoining islands, resulting in an accumulation of botanical and geological specimens, besides slaughtered bears and foxes, and quantities of drift-wood. On the 23rd the vessels parted company,—the *Tegethoff* steaming to the northward, and the *Isbyörn* endeavouring to push southward along the coast. On reaching the mouth of the Petchora, Count Wilczck and his friends left her to proceed on the return voyage to Tromsö, while they ascended the Petchora in small boats to Perm, and returned to Vienna by way of Moscow.

'The *Tegethoff* spent the winters of 1872 and 1873 in the Icy Sea, and made some discoveries of interest. It returned in safety in the summer of 1874.'

In lands further to the north than Nova Zembla there may be found moss, scurvy grass, and sorrel, but no trees.

In the year A.D. 879, when the first settlers in Iceland, under Ingulf their chief, went thither, they found very extensive forests in the valleys, which they penetrated with difficulty; and roots and stumps of large fir trees are, or were sixty years ago, to be seen in various parts, but now not a tree is to be found in the whole island, and only a few stunted birches and some low brush or underwood grows in the most sheltered situations.

Greenland is said to have gotten that name from an Icelander, Eric Raude, or Eric the Red, by whom it was discovered probably between A.D. 830 and 835, who, having explored the continent, returned in the third year to Iceland, where he boasted very much of the fertility of the land he had discovered, to which he gave this name, hoping, it is alleged, to induce others to follow him thither. A writer in the beginning of the present century speaks indeed of trees, but he tells: 'Those shrubs and trees, which in milder climates afford a comfortable shade to the wanderer, creep in this forlorn land under scattered rocks, to find shelter from their destroying enemies—storm, snow, and ice. This land, however, presents a series of plants which probably would not subsist in a milder climate; and in the interior of the inlets and firths many species previously unknown in other countries. Some of the new species are mentioned in the last number of the *Flora Danica*.* There are also other spots which boast the most luxuriant verdure, but they are only places in the neighbourhood of the Greenland houses, which have been improved for many years by the blood and fat of seals and other animals. There are also small hills on the uninhabited islands, where the birds build their nests, and, manuring the decomposing rocks, extort vegetation to their abode from the uncertain soil. These places, however, are but of rare occurrence, in proportion to the immense extent of the country. Innumerable cryptogamic plants, growing with great rapidity under snow and ice, supply the want of flourishing vegetation on the rocks and cliffs.'

According to reports brought by Baron Nordenskjold, the land, instead of being everywhere a green land, might with as much propriety as the land from which Eric was a fugitive, have been designated Iceland. On the 4th September 1883 he anchored in a fiord which had been newly visited by the Esquimaux, and where were found some remains of the Norman period. It was the first time since the fifteenth century that a vessel had succeeded in anchor-

---

* A classified list of plants, &c., found in Greenland is given in Brewster's *Edinburgh Encyclopædia*, vol. x., pp. 494-496.—J. C. B.

ing on the east coast of Greenland south of the Polar circle. The glaciers on the east coast are few, and of no great size; and the fiords are free from ice. Over the whole inland there is ice. There occur masses of fine dust, partly of cosmical origin, with the ice. An inland ice party started on the 4th of July from Auleitzwik Fiord. When they were 140 kilometres east of the glacial border, and 500 feet above the seal level, they were prevented by soft snow from proceeding with sledges. They sent the Laplanders further on snow-shoes (*skidor.*) These advanced 230 kilometers eastwards over a continual snow desert to a height of 7000 feet. The conditions for a snow-free interior consequently did not exist here; but this expedition, during which men have reached for the first time the interior of Greenland, has given important results as to the nature of the interior of an ice-covered continent. Baron Nordenskjold had believed that if the high mountainous ranges along the coast of Greenland were crossed, the valleys (if any) in the interior would probably be found covered with green vegetation, but it was found that after ascending to about 7000 feet above the sea level, the country was comparatively flat, and covered with snow and ice. The greatest cold experienced was 20 degs. below zero.

PART II.

FOREST EXPLOITATION.

CHAPTER I.

SARTAGE.

IN *Sartage* portions of the forest are burned down, and on the soil manured with the ashes different crops are cultivated for a few years, until its fertility being exhausted, it is abandoned, and another place is similarly treated.

In the narrative given by Mr Judrae of his journey through the forests of Olonetz and Archangel, mention is made of the practice of *Sartage* or *Rhoeden*, and *Svedjande*, as it is called in Finland—the last being a Swedish term introduced in connection with the domination of Sweden in Finland, previous to its being annexed to the Russian Empire. Mr Judrae, in the few sentences in which he has spoken of it, has said all, or almost all, that can be learned in regard to it from what is practised here. The literature of forest science in France supplies ample details in regard to it; and in reports of forest operations in India we are supplied with details in regard to the results and consequences of such a treatment of forests.

In a companion volume I have given details of the practice as followed in Finland, from which country, if not by the Finnish Karells inhabiting a large portion of the Government of Olonetz, it may have been introduced into this region. I consider it to be a practice of Asiatic origin, brought by the Finns from the East; and in the volume named details are given in regard to the practice as followed in India, Burmah, and Ceylon, with discussions which have taken place in regard to the advantages of this mode of exploitation under different conditions.

Both at Vosnesenya and at Petrozavodsk I heard of *Sartage* being practised frequently, and in different parts of the Government.

On this subject Mr Judrue says:—' Reading the reports in the Government office of the Imperial Domaines, one is arrested involuntarily at a place which treats of unauthorised fellings carried on without leave or sanction.

' According to these reports the population of the Government consists almost exclusively of those who were Crown serfs and their children, whose requirements of wood for fuel and building are sufficiently met by the allotments made to them annually from the forests; but these people for a long time back have been possessed with the idea that woods are of no pecuniary value, and they destroy them recklessly. When the annual allotment happens to be less than they think they require for building material —for it may be fancy erections which they do not require —they frequently go off to the woods and cut what they want without ever applying for permission to do so. And then the question comes up, Is it possible for the people to acquire at the present time any adequate idea of the necessity which there is for the conservation of the forests and the exploitation of them in a rational or scientific way? Let any one realise the case. Around all of these villages, even the smallest of them, there are forests of which the eye can see no end, they appear to be interminable; and there are depths of them to which the foot of man has never penetrated. The extent of these forests is such that to the peasantry they seem inexhaustible; while, on the other hand, the severity of the climate, the unproductiveness of the soil, and the poverty of the people are such as to seem to call upon every one to find out for himself with a hatchet in his hand any means of improving his condition.

'The natural condition of the country could not have called forth or exercised upon the people an effect more to be deplored.

'The peasantry here look upon wood as being in common with earth and air, fire and water, one of the elements, and as equally free to all persons; and they consequently consider that they are free to use it without stint or limit, as one of the free gifts of nature. This state of things, originating, as I have intimated, from the physical condition of the country, can only be changed or destroyed by the great change-producer, time; and the reports of the consequent destruction of the forests embrace numerous details of the extension in the country of the practice of *Sartage* and *Roeden,* or *Svedja*. This system of felling is very frequently met with; but if we enter into the circumstances of the case, considering, on the one hand, the condition of the agricultural economy of the people, together with the paucity of labourers and the lack of manures, and the circumstances that the temporary culture of the fields which is thus effected supplies the only means of support to man, and, on the other hand, the great extent of the forests and the difficulty of maintaining an efficient watch over them by wardens or forest watchmen with a great extent of forest entrusted to their care, we cannot condemn the Forest Administration for not adopting effectual measures to prevent altogether this unauthorised felling of trees in the forest.

'This unauthorised felling is the primary form taken by agriculture—the first step taken towards the development of rural economy. We hope in process of time to get beyond this; but to put it down by force would not be a rational course of procedure. The Northern peasant not having productive ground near his residence, nor means to improve it if he had, goes into the depths of the forest, burns down trees, and cultures his temporary field for two or three years, or so long as its powers of fertile production is not exhausted—the fertility being produced by the ashes and cinders of the burnt trees. The persuasion of the peasant as to the perfect legality of such a procedure is such, that it is very doubtful whether any general measure of repression at present could remedy the evil. In order fully to understand the economic condition of this region we must go back some fifty years or so, and look at things with other eyes. I consider that this unauthorised felling originally was legal and reasonable—suitable for the place where the forests are very dense; but as a principle it admits of some formal limitation. And this, according to these reports, appears to have been attempted in the Government of Olonetz in 1867. Of the system of operations carried on by this people, it is said the first settlers in the country were satisfied with small plots of ground of easy cultivation, but as they increased in number they were obliged to have recourse to land which was more fertile indeed, but marshy or covered with forests, and requiring labour to prepare it for culture, and care and thought. Cultivation such as may be seen in civilised communities was not attainable by these people, were it only for their want of agricultural implements and manure. In the same book, on the page following, it is stated, "In these virgin soils, previously covered with forest or bush, the produce of rye in the first year was *ten*fold—frequently *twelve*fold; and there were places—generally places where there had been old dense high forests—in which the produce was *fifty*fold, and in the second year the produce was from ten to fifteen fold."'

CHAPTER II.

JARDINAGE.

SARTAGE, which has been noticed in the preceding chapter, can scarcely be called with propriety *Forest Exploitation;* with more manifest propriety it may be called *Exploitation of Forest Land;* but the utilisation of the ashes obtained, and the utilisation of the forest trees to produce these ashes, justify my treating of it as I have done. With *Jardinage* it is otherwise, in as much as it is employed primarily as a means of utilising the trees produced, irrespective of the ground on which they were produced.

The designation is given in France to a treatment of forests prevalent everywhere, according to which a man seeks out and fells the tree which he thinks will serve his purpose, whatever that purpose may be, leaving the others standing, if they do not happen to be crushed by the fall of his tree, or stand in the way of his getting it brought out from the forest.

This method of exploitation gradually exhausts the forest of all trees yielding large timber, as does the practice of the gardener, from which the designation given to it has been derived, exhaust the bed of leeks, onions, turnips, or carrots, gathering one here, another there, as they come to maturity. Others have testified what they have seen of this effect in different countries in Europe and elsewhere. In a volume entitled *Hydrology of South Africa*[*] I have

---

[*] *Hydrology of South Africa;* or, details of the former hydrographic condition of the Cape of Good Hope, and of causes of its present aridity, with suggestions of appropriate remedies for this aridity.—In which the desiccation of South Africa, from pre-Adamic times to the present day, is traced by indications supplied by geological forma-

given the following account of what I have seen, not in one forest alone, but in many widely dispersed over the colony of the Cape of Good Hope, supplying illustrations of the first, the second, and the final stages of the devastation thus occasioned:—

Under a system of forest management which, borrowing a term employed in works on forest science in France, I may call primitive *Jardinage*, the forests in the colony have been long gradually disappearing. The system followed was to cut down trees such as might be required, leaving others standing, but doing nothing to promote their growth, or to replace those which were removed.

I have before me a chart of the forests of the Tzizi-Kamma. From information supplied to me by Captain Harrison, the Conservator of forests in the district, I have gathered the following particulars, which I give, as illustrative of what I may call the first stage of the work of destruction under the treatment which I have called primitive *Jardinage*.

On the west bank of Storm River there is—or was at that time—a piece of what may be described as virgin forest, in which operations were begun about ten years ago. On the east bank of that river is a patch of scrub destitute of timber.

Below this is a large piece of ground in two divisions, which is mostly private property, and in which the Crown property had been denuded of timber previous to Captain Harrison entering on his duties as conservator of forests in the district.

Continuous with this, and at the mouth of the river, is a patch in which wood-cutting has been actively carried

---

tions, by the physical geography or general contour of the country, and by arborescent productions in the interior, with results confirmatory of the opinion that the appropriate remedies are irrigation, arboriculture, and an improved forest economy; or the erection of dams to prevent the escape of a portion of the rainfall to the sea,—the abandonment or restriction of the herbage and bush in connection with pastoral and agricultural operations,—the conservation and extension of existing forests,—and the adoption of measures similar to the *rébois ment* and *gazonnement* carried out in France, with a view to prevent the formation of torrents and the destruction of property occasioned by them.—London : C. Kegan Paul & Co. 1876.

on for years, and in which timber is consequently becoming scarcer, but waggon-wood is still plentiful.

A little to the east of this is a patch which still contains some very large yellow-wood trees; and half-way between this and the mouth of the Faure River, which is still further to the east, is a large patch from which an immense quantity of timber has been cut out of late years, and in which the work is now going on daily.

In the upper district of the Faure River, skirting its east bank, is a patch, the timber of which has been nearly exhausted, but in which there are an immense number of young trees. And half-way towards its mouth is a patch which has been nearly destroyed by fire. It is a patch of Kuerboom—*Virgillia capensis;* and there is valuable timber in it.

A small patch skirted by the Kruis River on the east, has a few yellow-woods close to the river; but the other timber has been cut out.

To the north of this, near the source of a tributary of the Kruis River, is a larger patch, from which the timber has been cut out, but in which a few young trees are growing up.

Below the confluence of this tributary of the Kruis River, traversed by another shorter tributary stream, is a patch in which stinkwood is becoming scarce, but in which yellow-wood and waggon-wood are plentiful.

Continuous with this, lying in the fork formed by the confluence of the Kruis and Eland's River, is a patch which was formerly private property, but which is now the property of the Crown, and which contains valuable timber at its lower extremity. Near the confluence of the rivers continuous with this, but on the eastern bank of the Eland's River, and extending towards its source, is a patch of valuable timber of all kinds, but the trees are growing in deep kloofs. Below this, and continuous with it to the banks of the Stinkwood River, and the confluence of this and the Eland's River, is a large portion of the same patch, containing valuable timber of all sorts, of more easy access.

And continuous with this, on the eastern bank of the Stinkwood River, is Robbe Hoek, in which is sound valuable timber, but it is difficult of access, in consequence of its growing in deep kloofs. In this patch waggon-wood is plentiful.

Above this are three small patches in which no valuable timber has been left uncut, but in which a few young trees are growing up. Still higher, skirted on the east by the Witte-els River, and traversed in part by the upper bed of the Stinkwood River, is the Witte-els Bush, near to which is the residence of the conservator. It abounds in witte-els and contains good waggon-wood, but the stinkwood and yellow woods have been nearly cut out.

On the south or seaward side of the Eerste River, where it follows a course parallel with the Eland, is a large portion of an extensive patch traversed by that river, in which there is plenty of waggon-wood and some very large yellow wood trees, but little stinkwood has been left in it.

On the north bank of this river, and on the same side of the river, where it takes a southerly course, are three patches, from which all old timber has been cut, excepting such as is not generally used, and these patches are now closed to allow young timber to grow.

Such is the first stage of the work of destruction under the treatment which I have designated primitive *Jardinage*, here arrested, it is to be hoped, by the judicious measures adopted by Captain Harrison. But the progress of the work can be traced a little further in an adjoining district in regard to one Crown forest, in regard to which the forest warder wrote to me some time since: 'I would suggest that Government should, without delay, get this portion surveyed, as ——— and ——— are appropriating the forest to themselves. No licences are exhibited, and to my knowledge, as much as £750 worth of timber has been removed within the last ten years, while for the cutting of timber out of the said forest I have only issued two licences [each for the removal of a single load.] The same amount of value in timber has been destroyed through the reckless behaviour of these individuals, and those in their employment, igniting the grass, which has caused fearful destruction. There are a few other small patches and stripes of bush; but, comparatively speaking, they are nothing, only adapted for fuel; most of the valuable timber has been removed, and by fire greatly destroyed. The great evils are men cutting without licences, and grass fires.

'To my knowledge, there is on an average 40 loads of *fuel, poles,* and *spars* removed weekly to Port Elizabeth from the forests between the Gamptoos and the Van Staden Rivers, for cutting timber for which I have never issued one licence for the benefit of Government. I feel convinced that it all comes from the Crown forest; but as it is a case of disputed boundary and licences, I am not empowered to move in the matter. If this state of things continues much longer, the whole of the forest will be eradicated and destroyed.'

Such an issue as is thus indicated may be considered the second stage of the destruction of forests under primitive *Jardinage*, the conversion of forests into *bush*. In Krakakamma, between the Zitzikamma and Port Elizabeth, there is a good deal of arborescent vegetation, but it can scarcely be reckoned forest; the same may be said of the Kadouw Bush, between Port Elizabeth and Grahamstown; and such, I am informed, is the present condition of what within the last thirty years was an extensive forest in the valley of the Kowie, in the neighbourhood of Bathurst: the old timber having been destroyed, but not replaced, the forest character has been lost.

But this second stage of the progress of the work of destruction is not unfrequently succeeded by a third, in which even the arborescent bush may disappear. From more than one of my correspondents I have heard of the mountainous country around Somerset having abounded in forest trees of various kinds—Yellow-wood (*Podocarpus*), Iron-wood (*Olea*), Assegai-wood (*Curtisia*), but all of these are fast disappearing. Mr Leonard, of Somerset, in reply to a query issued from the Colonial Office in 1804, having remarked that the Yellow-wood tree forms a much less conspicuous element in the scenery than his memory pictured it doing some four-and twenty-years before, goes on to say,—' Of other forest trees there used to be an abundant supply in the forest that skirts our mountain here, but the large demand that rules in an age of bullock waggons for disselbooms and other waggon wood, is sure to clear out any but an inexhaustible supply of Assegai and Iron-wood trees, while the durability possessed by the olive post soon marked it out for the woodman's axe, in procuring timber for the ever memorable Hartebest house of the first pioneers; and subsequently the same durability in the nature of the wood caused the continuous destruction of the tree for fencing stakes, when advancing civilisation demanded and gave way to buildings of brick and stone.

'Yellow-wood trees of any size, as well as Assegai, Olive, and Iron-wood trees are now becoming so scarce here that we may easily predict the speedy extirpation of them from amongst our natural productions; and, unless human care and culture produce specimens, when those of the kloof and the rivulet have disappeared, the next generation will have to refer to some some botanical collection to see what they are like.'

About the same time the late Rev. J. W. Pears, the minister of the Dutch Reformed Church at Somerset, previously professor in the South African College, Capetown, writing to me on another subject, said:—' When I came to the frontier 38 years ago there was grass everywhere in abundance, in the plains sweet, and in the mountains sour; and this, sometimes five or six feet high; now none, excepting near rivers or on the tops of mountains, is to be found. Formerly, also, the mountains were unoccupied, as no one chose to pay for them; the herbage was abundant; and the moisture was long detained, so that all the little streams continued to flow through the whole year. Now these mountains were all occupied, and generally burned annually, and the consequence is that the water has failed. For instance, the mountain behind my house, which rises to the height of 1,756 feet, was covered with high grass and thousands of beautiful bulbous flowering plants and shrubs, and its whole face and offshoots adorned with Yellow-wood or other valuable trees; now these are all gone; not a Yellow-wood or other tree worth anything is left, and only a useless growth of bushes occupy their place, and the consequence is that a stream that supplied my garden and some others, runs now only after rain. The whole face of the mountain, if planted with oak, firs, and other useful timbers, might not only be valuable, but again it might protect the water. But almost every year, by the idle and reckless, the mountain is fired, and all is destroyed. It is now burning fiercely. In the kloof there still stand the charred stumps of large Yellow-wood trees.'

Such appear to be the only remains of the forests once flourishing in the neighbourhood of Somerset.

This may be considered as a third stage of the destruction of forests—the final—in which they entirely disappear. And to this those spoken of as being destroyed in the vicinity of the Gamptoos River are likely soon to come. I am informed that 'the whole of the Crown Forest Reserve and vacant land in the ward of Van Staden's River, which comprises also the Field Cornetzy and ward of Eland's River, is to be disposed of on a twenty-one years' lease; other portions, not of great extent and value, are to be annexed to the properties adjoining them; and the office of Forest Ranger is to be abolished.'

Similar results have been seen by others elsewhere.

In a paper by Lady Verney, in the *Contemporary Review*, I find the following statement of a generally accepted fact: 'The question of the supply of timber for the future is all over the world becoming very serious; the sources are gradually exhausted, while scarcely anything is done to repair the waste, except by England and in parts of Germany. In India the small cultivators cut down the trees wherever they can, and, of course, never plant, and the destruction of the forests has greatly injured the rainfall, dew-moisture, and supply of wood in the country, while the peasants are burning manure of their cattle for lack of better fuel, instead of putting it on the land. Government has now been obliged to interfere, both for the protection of forests and to plant fresh trees. In America along the whole line where cultivation encroaches on the backwoods, the trees are recklessly destroyed, even burnt down, and no steps are taken to ensure future supplies of timber in place of that which is so rapidly disappearing. What is sent to Europe comes every year from a greater distance inland.' And so is it here.

Mr Judrae, in his account of his journey through the forests in the Government of Olonetz, makes mention once and again of the owners of saw-mills which he visited complaining that the exploitation of the forests had become unremunerative. This is attributable not to any falling off in the demand for timber, or to reduction in the prices obtainable, but to the increased expense in procuring timber, while the other conditions remained the same. And this increased expense is attributable not to a rise in wages, but to the greater distance from which trees must be brought to the mill in consequence of the exhaustion of these in the immediate vicinity.

In accordance with the complaints reported by Mr Judrae, by a gentleman who had for years been engaged in another of the departments of the exploitations, I was informed that no trees were allowed to be felled within six versts, or four miles of the river; and that with a view to the conservation of the river, as well as the conservation of the forests, no tree was felled but such as would yield a trunk free from branch or bend, 37 feet long and 7 vershoks, or 12 inches, in diameter at its upper extremity, apparently free from shake and from decay, and such as must be felled in order to the removal of the tree required.

The trees were felled not more than 28 inches high, the trunk of the required dimensions alone was removed, the stump and the head were left to rot.

The trunks are subsequently examined and marked. They are dragged over the snow, and launched on a streamlet or river, and floated to a locality lower down the river, where they are collected by a weir in a receptacle calculated to hold from 1000 to 2000 trunks. When this is full they are dragged out along an inclined plane of timber, by a rope attached by the ends to a post in the land beyond, the lengths being passed over the two ends of the trunk, and attached by their extremities to a splinter bow, and by horse-power they are rolled up to the land.

By steam-power and water-power they are sawn up into planks and beams by sets of swing saws. A fire is constantly burning to consume the *débris;* and I have otherwise learned that the sawdust is sent off to peat bogs and such like places to prevent its accumulation. It has been found practicable elsewhere to use sawdust as fuel in steam producing furnaces, but with the surplus of *débris* this is there unnecessary.

In this case, and in many others, perhaps on all in which the exploitation is carried on upon a large scale, and under proper supervision by officials in the forest service of the Government, the exploitation, though carried out in accordance with the practice of *Jardinage*, is executed systematically and with some regard to the requirements of the future. But in many cases—I may almost say, in most cases—this is done recklessly and without any consideration of what may be required in the immediate future or in that which is more remote, and the information I have received from others is in accordance with my own observations.

By a forest inspector I was informed that there are whole districts in which it cannot be said that any systematic management whatever of forests is observed—Unregulated *Jardinage* is the only description which can be given

H

of what is done in the felling of trees; and that in Siberia there is not even a system of taxation or of charge for trees that are felled: all which is in accordance with what I have heard from others. One of my correspondents who had occasion to travel extensively every year in the interior told me it had frequently happened, when he asked in regard to forests through which he was passing, to whom did they belong, he was told they were free, by which he understood that any one might fell trees without let or hindrance. They may have been communal forests; but it happened too often and too far from villages to give countenance to the supposition that such was the case.

I have spoken of exploitation by *Jardinage* as being destructive of forests, and have in illustration cited what I have seen of its effects at the Cape of Good Hope. Similar effects have been witnessed both in Germany and in France. It was this which gave occasion for Colbert's oft-cited saying, *France perira faute de bois!* and for the celebrated Ordinance of 1669, known to, or known of by, most students of *Forst-Wissenschaft*, or Forest Science, on the Continent of Europe. And even in Russia, where forests cover regions which in extent appear interminable, it is seen to be only a question of time when forests will disappear, unless the measures now being adopted by the Forest Administration and patriotic landholders in various parts, or others, to which these may give rise, shall avert the evil. An approximate estimate has been formed of the cubic measurement of the annual growth or increase of wood in Russia, and an approximate estimate has been formed of the cubic measurement of the wood annually consumed as firewood, building material, raw material of various manufactures, articles of export trade, &c.; and this has been found to be so far in excess of the former, that, but for the remedial measures referred to—the consumption continuing undiminished, while the production was every year becoming less—in some 200 years Russia must be divested of her forests.

JARDINAGE.

Yet the system has its advantages, in view of certain results which may be sought. It is not a mode of exploitation producing in all places and in all circumstances unmixed evil—evil, only evil, and that continually. The object I have set before me in sitting down to prepare the following report is not to condemn *Jardinage*, neither is it to commend or to justify it, but to supply information in regard to its details as carried out in the North of Russia, leaving to my readers to make what use they may desire, or be able to effect of the information given.

The object aimed at by the most advanced forest management of the day is to secure by the operations adopted a sustained production of wood, a progressive amelioration of the condition of the forests, and a continuous material reproduction of the woods; and this is called for in Germany and in France.

But there are cases in which a great quantity of wood is suddenly called for—cases in which a continuous supply of fuel or small wood for other purposes is desired—and cases in which the production of wood of a given bulk at a period more or less remote is required. The treatment given to a forest must be different in each of these cases.

There are also cases in which it is not wood, but the money for which wood may be sold, that is wanted—money coming in in instalments over a period more or less protracted, or money required at once; and measures must be adopted accordingly.

Again there are cases in which it is amenity and shelter, or a covert for game, irrespective, it may be, of all besides, or it may be along with one or more of the objects specified; and again this must determine the course of action followed.

And yet again it may be none of all these things which is desired, but the ground upon which the forest grows which is wanted for horticultural or agricultural purposes—or the clearing away of the forest for climatic effect, has been resolved on; and again a particular course of action is called for, and this may be modified according as it may be sought to secure along with this a supply of wood for use or for sale.

It must be apparent that none of these advantages are secured by *Jardinage*. But on the other hand, there may be found, in climatic changes and extended facilities for agricultural operations, for which such climatic changes would be favourable, compensation for the destruction of forests, resulting from this destructive mode of exploitation.

With regard to climatic effects, it may be stated that while some countries have suffered in climate, as have Spain and South Africa, from paucity of woods, and an unequal distribution of what there are, there are countries which suffer in climate from a superabundance of these, as do Finland and the North of Russia. The humidity and shade desiderated in the former, are in the latter in excess. Though Russia is said to have a large portion of her area forest lands, these are found chiefly in the Northern Zone. Observations collected by the Agricultural Department of the United States of America tend to show that to secure the greatest climatic benefit from forests, the forests and the arable land should bear a definite proportion to each other, varying with conditions, not only over extensive areas, but over limited divisions of the country; and students of Forest Science in Russia can contemplate with calmness the possible disappearance of forests over extensive areas of the Governments of Archangel and Olonetz, provided adequate measures be adopted for the conservation of forests in the midland Governments of the Empire, and for the extension of these by sylviculture in the south.

In the North of Russia *Jardinage* may lead as certainly to the destruction of forests as *Sartage*, or the burning down of the trees with a view to rearing cereals for two or for three years on the ground fertilised with their ashes; but this, which in some circumstances would entail a curse, may there bring a blessing; and meanwhile attention is given to considering only how the mode of exploitation followed may be carried on as advantageously as possible.

CHAPTER III.

VIEWS ENTERTAINED IN RUSSIA IN REGARD TO DIFFERENT METHODS OF EXPLOITATION.

ACCORDING to a statement made by Mr Werekha, in a *Notice sur les Forets et leur Produits*, &c., prepared by a Special Commission charged with the collection of products of the forests and of rural industry for the International Exhibition at Vienna in 1873, 'in Russia the systematic felling of forests was formerly dependent on license or authority, and it is generally so even now. The clearing of forests by systematic fellings by sections goes no further back than to the time of Peter the First, and it is not yet sufficiently practised.

'The ancient *Jardinage* exists still in the greater number of forests; but this manner of arbitrary exploitation does not now satisfy the sale and requirements in many places, for some kinds of trees as necessarily predominate in the forests of the north and the north-east of Russia; and it gives way in the central and southern part of the Empire in proportion to the development of the demand over the whole extent of the forests to exploitation according to the system of management by regulated fellings.

'The abundance of forests in the north of Russia, and the little demand which there exists, are the cause why until now many of the forests situated in the northern countries cannot be exploited otherwise than by *Jardinage* to meet the limited requirements of commerce and local need.

'It was only in the latter half of the last century that they began to make special plans or charts of the forests and to prescribe the extent of felling which was to take place. According to these old plans, this was carried out by dividing the great extent of forests, even of timber forests, into narrow parallel zones across all the forest masses, and in number eqnal to the number of years of the revolution prescribed for exploitation.

'The physical inconveniences and the utter inequality of the produce of these fellings, according to *à tire et aire*, were the cause why these plans were carried out almost nowhere, and that the forest continued to be exploited arbitrarily by *Jardinage*. The scientific system of exploitation was not introduced into Russia and put in practice till 1841. At the present time over the whole of the Government forests there are 11,872,500 hectares subjected to a regulated exploitation, principally in the provinces of the south, of the centre, and of the south-west of Russia, where the forests have acquired a great importance, because their extent and their produce scarcely suffice to meet the local wants of the population. And the forests belonging to the mines and manufactories are all exploited according to a definite plan of management, and this is why one may reckon that in all these forests, which cover a space of 5,891,638 hectares, the fellings are in legalised proportion with the annual increase of the trees.

'On the forests belonging to the appanages there are 3,728,346 hectares which have been subjected to a regulated exploitation. In what relates to forests belonging to private proprietors there exists no organ which combines the technical data and statistics, and which contributes to establish a useful principle and practice. Yet of late the proprietors of forests, and especially the great proprietors, are earnestly desirous to submit their forests to a regular exploitation, and in acknowledgment of the advantage of it, have taken to administer their forests special officials who have previously received technical instruction on forest administration. It is even possible to cite some proprietors whose forests are managed rigorously, conformably to the rules of science, amongst others Prince Paskévitch, proprietor of many extensive forests in the Governments of Mohileff and of Riazan; Count Ouvaroff, proprietor of forests in the Governments of Minsk, Vladimer, and others; Prince Yousoupoff, Count Tolstoië, Count Strognoff, MM. Maltzoff, Demidoff, Schatiloff, Scheremetief, Countess Ribeaupierre, Count Apraxin, Baron Korff, and some others. Even in forests belonging to the communes of peasants there begin to show themselves here and there some rare attempts at reasonable management. Amongst the forests belonging to towns those which belong to Riga and to Pernau are well managed. But in by far the majority of forests belonging to private proprietors they fell according to *Jardinage*, and here and there to the system of *à tire et aire*, and the system *à blanc étoc*, without any fixed plan of management, and solely according to the demand, or according to the want of money. The principles regulating the organisation of the management of the State Forests are not rigorously defined, but the better policy, proclaimed in 1841, has begun to prevail, that is to say, an exploitation designed to obtain the greatest material product, and that which will be most useful to the general interest. Elaborate plans of management determine the duration of the revolution of the fellings and the site of these, the estimate of the produce in volume and in value of the fellings of the first decade, the means of replenishing the timber forests, and the reproduction of copse woods, of reforesting of cleared spaces, void places, and vacant lands, and the local measures to be taken for the amelioration of the trees composing the forest most advantageous to its exploitation.

'These works are entrusted to Commissions of forest organisation, which, after having presented to the Forest Administration the general plan of a forest, with specification of the works of management for the first decade, proceed to the working out of analogous plans in another forest; and such a Commission towards the end of the first decade return to control the execution of the works of management of that first decade, and determine the special site for the operations of the second decade.

'The system of exploitation of the forests prevailing in

the central and southern portion of Russia consists in a succession of fellings making a clean sweep, contiguous to each other. The extensive replenishment of extensive surfaces denuded by such fellings through seed cast from the adjacent masses of forest is very rare, and one has tried to introduce into the exploitation of timber forests (which for the most part is very irregular) instead of a clean sweep, a reserve of trees for sowing it; but the number and the quality of the trees left standing as a reserve do not correspond with the rules of forest art, nor do they accomplish the end designed. The execution of contiguous fellings frequently renders it necessary to wait a long time till the effective sowing has been secured before beginning to fell the portion next adjoining. This delay in felling is often injurious to the quality, and to the product of ripe wood, or of trees too old and subject to deterioration; all of these inconveniencies have determined some foresters to take as their guide the scientific views given in foreign works of the last century, and to introduce the system of fellings by alternate bands. But these also have rarely given satisfactory results in the reproduction of the forests. The introduction of methods of natural re-sowing by successive fellings designed first to augment the production of the sowings, and then to protect from the wind the seedlings during the time necessary to their development without hindrance to their growth, and, in fine, the final felling is desirable. But this course of replenishing fellings, with the thinnings or periodical fellings of improvement, are not met with in Russia, excepting in rare circumstances; because the application of this the most rational method of replenishing the forest by natural and cheap sowings, encounters serious obstacles through the prevalent practice of giving to the purchaser the whole charge of felling and trimming in the forest.

'It is impossible to exaggerate the injurious influence which this usage exercises in Russia over the development of forest economy, and of sylviculture in general, for the superintendence of the woodman, during the felling of the trees, and during the trimming of the felled wood by the purchaser, who pays the woodman, will be always defective and insufficient. The forest wardens of different grades are powerless to remedy this, their influence and their authority over the woodmen is infinitesimally small—almost nothing—in all that relates to necessary cares for the good conservation and maintenance of the forests.

'It is impossible to subject to the conditions and requirements of forest economy him who exploits a forest which he hath bought with a view to felling as quickly as possible that he may be able to profit by the speedy return of his capital. The absence in Russia of the custom of conducting in the forests all felling and trimming by the local administrators or official foresters, or by the proprietors themselves, explains the extremely rare introduction of *coupes d'amélioration*, or periodical thinnings, so useful in the development of the growth of forests, the fertility of which cannot be obtained but by the strict and constant superintendence on the part of foresters, forest wardens, or proprietors conversant with all the details of forest economy. It is only by labours carried on thus that we can hope duly to culture the trees; and teach workmen selected from amongst the inhabitants of the vicinity to give themselves to the different departments of forest work; and to impart to them that interest and skill in the management of forests, without which the depredations in forests can never be diminished.

'In the Russian forests it is the natural wild reproduction by seed or by suckers and shoots which prevails; the artificial renewal by sowing or planting exists only in a small number of localities in which, through exceptional economic conditions, the management of forests takes a more intensified character.

'The plantation of new forests in localities altogether treeless is taking place chiefly in the steppes of Southern Russia, where, since 1842, the success of this enterprise has been secured by certain administrative measures. According to the reports of the Government Direction of forests, in the Governments of the South the artificial plantation has thus far progressed—

		Deciatines of Forest.	By Sowing.	By Planting.
1866	...	Cultivated 4148	2088	2060
1867	...	,, 2772	1372	1400
1868	...	,, 3007	1317	1690
1869	...	,, 3750	1447	2303
1870	...	,, 1230	255	975

'In what relates to the planting of forests on the properties of private individuals we possess very little information indeed; it is only known that with certain proprietors the plantation of forests has taken place on extents of land which are relatively considerable; with Count Ouvaroff in the Government of Moscow, to the extent of 700 deciatines; with M. Schatiloff, in the Government of Tula, with M. Skarjinsky, in the Government of Cherson, to the extent of 500 deciatines; in the Colony of Mennonites, in the Government of Taurida, and with some other proprietors.

'Works of improvements which consist in constructing or improving forest ways for the leading out of fellings, the drying of marshes, the surrounding the borders of forests with ditches or quick-hedges, and the purchasing up in Courland of forest servitudes have only taken place in what are relatively very restricted proportions.

'Conformably to the natural geographical distribution of the principal forest trees in Russia (in the north they are resinous woods which predominate, in the south they are leaf-bearing trees), the management of the forests has taken also two different principal forms. In the north, and in the northern portions of Central Russia, there predominate lofty forests with reproduction by seed; in the south, and in the central portion of Central Russia, coppice woods predominate, with reproduction by shoots and suckers. The success of this last system of exploitation meets with a serious obstacle in the want of a strict organisation of the use of pasturage in the forests. Composite exploitation, or that of coppice mixed with timber trees, has extended without preconcerted plan, and solely by isolated cases, into the western provinces of the region of black soil.

'Besides these principal forms of forest management, there is met with in the Russian forests the application of some other varieties of exploitation relating rather to the soil than to the forests. *Sartage*, not only of coppice, but frequently of tolerably high perches of resinous trees, for the most part without any system, exists in the Governments situated in the north and in the north-east, but it tends from day to day to disappear. *Sartage** consists in this, that on fields exhausted by the culture of flax or of cereals, the poor soil, by its nature where impoverished by exhaustive culture, remains fallow during a very long time; but in the long run it covers itself with an arborescent vegetation which, penetrating with its roots into the unexhausted subsoil, makes rapid increase. When nature has accomplished this process, the peasants, after having cut the wood, burn it on the spot as they have no sale for it, sow flax or corn for some years on the soil enriched by the ashes, and when indications of exhaustion appear on the ground so treated, they leave it anew in fallow. This ancient mode of culture is still in use in the southern and western countries of Europe. Traces of a more regular organisation of alternative management consists in utilising the soil, now as fields of labour, now as soil covered with forest trees, and may be seen in some spots in the central portion of Russia, for example in the district of Melenkoff, in the Government of Vladinir, in the district of Mojaisk, in the Government of Moscow, where, after the removal of the fellings, the soil of these is put under culture with rye, oats, buckwheat, and other plants used in domestic economy during two or three years, and after

---

* *Sartage* has been brought under consideration in a previous chapter [ante p. 85], and the advantages and disadvantages of this mode of exploitation have been described in a companion volume entitled *Finland: Its Forests and Forest Management.*

that time it is re-wooded, most frequently by means of sowing, and more rarely by help of planting; besides, the expenses occasioned by the re-foresting are generally covered and even more than covered by the proceeds obtained from the agricultural cultivation of the soil in the interval.

'The system of culture by pollarding, which consists in cutting periodically the shoots from the stems of the willow and other trees of underwood, is not met with but in exceptional cases, and more especially on spots exposed to inundation, as for example, on the banks of rivers and of ponds in the western and southern provinces. The system of periodical pruning of lateral branches may also be met with in the forests of the southern Governments, more especially near villages, but it is only rarely applied, and that in isolated places.'

In a notice of Russian Forests in the *Journal of Forestry*, vol. ii., April, 1879, p. 881, it is said:—'It is stated that the newest financial project of Russia is in the form of a concession for a term of years of all State forests to a joint stock company, which will have the exclusive right of cutting and selling timber from these sources in return for a stipulated royalty, to be paid annually to the Government. Among other conditions of the concession the company is to be held bound to replant wood lands that have been already denuded by injudicious forestry or by theft, and to conduct their own felling operations with due regard to the interests of the future.'

CHAPTER IV.

EXPORT TIMBER TRADE.

It is in the Governments of Olonetz and Archangel that the export timber trade of Russia has attained its greatest developments, and one of the most important marts is the town of Onega, which is situated on the river of that name at its embouchure into the Gulf of Onega on the White Sea. It flows from Lake Latcha, taking its departure not far from Kargopol. The district was aforetime annexed to the Republic of Novogarod, which granted permission to Dutch and other merchants to cut wood and dig iron and mica in the vicinity of the lake.

Lake Latcha is connected with Lake Woshe a little further to the south. The river is a broad stream some four hundred miles in length, flowing through some fine scenery, and having its course varied by some fine rapids.

For many years a considerable exportation of timber has been carried on by the English Timber Company, or the Onega Timber Company, as it is generally called, which has held a concession for the cutting and exportation of timber from the district.

The most extensive timber operations in this district, and they may be considered the most extensive operations of the kind carried on by any one party in the empire, are those of this company, which is understood to be a company of British capitalists engaged extensively, if not exclusively, in the timber trade. From a gentleman well acquainted with the local administration I received the following information:—

'Most of the trees cut down for timber grow within a

limited distance from some stream, or in some locality whence they can easily be transported to some river or stream. In these cold sterile countries which cannot grow grain, timber trees grow to perfection; and the long winter facilitates operations, as during that season snow covers the ground, and by the aid of sledges the peasant can convey the timber to the banks of the river over districts where no wheel can pass. When the snow melts the rivers rise, and the timber is thus floated down at small expense to the mills. Forests far away from rivers are not valuable, the expense of the carriage of timber being great. The value of a fir tree averages 3s, the cost of cutting is about 1s, the floating to St Petersburg will be about 1s 9d. The branches are used for firewood, and the bark is cleared away, buried, or burned. The supply has hitherto been maintained by the abundance of the timber in the woods; but latterly they have been going deeper and deeper into the forests, and further and further and further from the navigable streams; and there can be no doubt that in course of years the supply must decrease unless certain restrictions are established.

'In our contract with the Government it is stipulated that the timber shall each year be cut only in the district pointed out by the forest officers. By this arrangement the forests are not destroyed, but thinned out periodically. With us large trees do not suit, as in them the centre is to some extent decayed. We generally cut down trees measuring from four feet to five feet in circumference.

'The Company have two saw-mills on the Ponga and one on the Onga, both tributaries of the Onega.'

From another gentleman I received the following more detailed information in 1874:—

'The Onega Wood Company have made a contract with the Russian Government to the effect that it has the right during the space of twenty years to fell not less than 60,000 and not more than 200,000 trees in each year. The *datcha* (districts) in which the Company may cut their trees extend from the northern part of the Lake Latcha along the river Onega, which flows out of this lake, to the small village called Porog, where booms are placed across the river to direct aright the floating timber. This is the first district and it is that from which the largest and best trees are obtained. The second district, Podporog as it is called, is from Porog to the town of Onega. This includes all the tributaries of the river Onega from its source to its mouth. Kargopol is in the Olonetz Government, 61° N. and $56\frac{1}{2}$° E. of Greenwich. It is prohibited to cut down any tree within six versts of the banks of the river Onega or any of its tributaries; this is in order to prevent the earth along the banks of the rivers crumbling away, and thus the rivers to shift their water-courses.

'The Onega Wood Company have to pay what is called hand-money, in a prepayment of 55 kopecs per tree for all intended to be felled. Thus, if they intended to fell during the following winter, for this always takes place by law when the sap has gone into the roots and there is little vegetable vital action in the tree, say 100,000 trees, they would have to pay 55,000 roubles.

'The Company make contracts with the peasants of the various villages in their district to fell so many trees, this being done not with a single individual, but with the whole family, and this family have to get the trees from any place in which they can find them beyond the prescribed distance of six versts from the river, and to bring the log to its banks: the only proviso made by the Company being that the river, or rather rivulet, shall be able, when the snows melt in the spring and the ice is gone, to float the timber.

'The Government make the additional proviso that the tree shall be felled not higher from the root than one arshine, or 28 inches, under penalty of twice the value of the tree. The log is to be 16 arshines long, and by the diameter at the smaller end the size of the tree is determined. The Company also make a stipulation with the peasants that the log shall not have root-rot or *lip*, a place where the bark has not closed over the wood, this being considered a sign of rot; also, that there shall be no branches within these 16 arshines, or 37 feet 4 inches.

'The *lip* may generally be seen wherever the root-rot is discovered by the sound the tree gives out when it is struck with the back of the axe, which, when all is right, should be a perfectly sound, hard, and somewhat deep note. The *lip* is situated generally high up the tree, and it is often the effect of a branch having died and fallen off, leaving a bare place.

'These logs are carefully bracked, or approved and assorted, by the bracker, who measures the breadth at the top, marks the number of vershocks which it is in diameter by certain chops on the edge, and then, by means of a hammer cut with his initials, his initials are stamped in at the top and bottom of the log.

'When a ten-vershock tree, or twelve inches in diameter, is felled, sometimes another length can be taken off one of seven vershocks, twelve and a half inches in diameter at top, from the same tree. This, however, is seldom done, and it does not pay, for neither is the bottom length good nor the top of much value. The bottom length is too broad grained and very liable to rot, the top is generally full of knots.

'In the spring the *Splavtchick*, or floater, comes with his *artel*, or company of men, and pushes the trees into the rivers, and sees that they go down to the place destined for their being counted by the Government official. If by chance a tree runs ashore, he is bound to get it off. He is provided with a boat-hook, and if it happens that he is on one side of the river, and the tree is stranded on the other, he gets on to a log which may be floating by him, and though standing on this round surface in the water, he is able by means of his boat-hook to paddle across the river without losing his balance.

'When the logs come to Podporog, they are caught by the booms stretched across the river; here they are counted by a simple process of sending a quantity into a space which will just contain one or two thousand. The ingress is then closed, the trees are bound together in *gonki*, floats, of from 300 to 1000, by means of branches and ropes, and these prams are then carefully floated down to Anda, the highest water-mills belonging to the Company, or straight down to Ponga, opposite the town of Onega, where the steam saw-mills of the Company are situated. I hear that Anda has now also a steam saw-mill, so that the Onega Wood Company has three steam saw-mills at work.

'*Porog* is the Russian for rapid, *Podporog* signifies below the rapids; up to this village the river is not influenced by the tide; lower down, or below the rapids, it is.

'The trees are then hauled out of the water. This process is done by the engine itself, or, when the trees are to be stacked, by a couple of horses. The ends of a long rope are attached to the stack, to the centre of this rope a rope is fastened, this loop is then passed round the tree, being slipped on over the trees when they are in the water by the Vikatchick; the horses are then driven forward, and the tree rolls up the incline, held at its two ends by the rope.

'The sawing is a process of some nicety, and requires great judgment. In an accompanying diagram I have drawn the various dimensions of trees at the thin end, seven to ten vershocks. The seven vershock trees produce only French boards—these are only exported to France; few in comparison go to England. The 11-inch boards are more in demand in England. I said it is a matter of some nicety to saw a tree up well. The heart-shake must lie in one of the boards. If the *mitick* (heart-shake) runs parallel through the tree well and good; if, however, it has a twist and is at the other end at right angles, that log is of no use as timber, and can only be used for building. If a board were sawn out of it it would fall to pieces. The sawyer takes pains to choose his trees much of a size, and so to arrange them as to get as many saws through the log as he can.

'The slab is first cut into shape, and the parts thus removed are thrown away and burnt, or given to the poor people. This means of getting rid of wood, however, is not sufficient,—a perpetual bonfire is kept near the works to destroy these, and there are destroyed thousands of pounds worth of wood. If that wood were only in England, or if freight were not so very high, it would not be wasted. The two sides are then cut off from each slab by the circular saw, thus leaving two boards of 9 by 3 in., and two of 9 by 1¼ in., obtained from a seven vershock tree, are 12 in. in diameter at the smallest end.

'Formerly in the water-mills there were no circular saws, so these side pieces were first sawn off, and then the boards were cut. Now, however, it is different. The sawing by steam takes about twelve minutes, that by water fifteen to twenty minutes, to do an equal amount of work.

'The timber trade is not so destructive to the forests as is the firewood trade, for, of course, the finest trees are only chosen, the rest are left. They cast their seeds, and a young growth is always coming up, which in the course of years will produce fine trees adapted for sawing up into planks. The firewood trade, on the contrary, lays bare whole districts of beautiful wood country, and exposes the land to the effects of the cold winds, and has a material influence in changing the climate and soil of the country. The wholesale destruction of forests has begun to be felt by the people of St. Petersburg very severely. Formerly wood cost 2 roubles 90 kopecs per fathom, seven feet square, now it costs from 4 roubles 80 kopecs to 5 roubles a fathom.

'One of the principal reasons why the seven to eight and a half vershock trees are best is, that not only are the knots in the boards all sound and good, but that the grain of the wood is far closer, and the inner or heart-wood is greater in proportion than the outer or laburnum.

This laburnum is more easily affected by the atmosphere, and where it is exposed to water it much sooner rots.

'When the boards are sent out from the mills they are stacked so as to allow as much air to get to them as possible. One layer is put on the flat, the next is put on the edge, and so on; thus the air gets at them, and they season and dry. In the spring they are all bracked or assorted, 1st, 2nd, and 3rd sort, battens and deal ends. The first sort of deal ends only are exported, the other sorts are retained for home use. Much judgment and a quick eye is required, for often by cutting a piece off the board $\frac{3}{4}$ may pass as 1st sort, and the other make a 2nd sort deal end; or perhaps by cutting it in half, half may pass as No. 2 sort, and half as No. 3 sort. Sometimes by merely cutting a few inches out, the two lengths are good enough to pass into the 1st sort. In the spring all the ends of the deals are cut, this takes off the rough end left by the felling axe, and as this process is only done after the boards are bracked, it ensures that each board goes through the bracker's hand. He writes the sort on each with a piece of red chalk, makes a line where it is to be cut, and a cross on the place to be thrown away, or a D if into a deal end, and 1, 2, or 3, if into different sorts.

'They are then stacked into close stacks, all the boards on the flat, and quite close together, or with merely an inch between every two. They are then taken down the river in crafts, and go down to a place called Ki Ostroff, a little island at the mouth of the river, I might almost say of the bay, some twelve miles off. Here they are again stacked, and when the ships come they are loaded into crafts, and are taken by the tug to the anchorage grounds some three miles off. The loading is difficult, as the place is open to the winds from the N.NW. Even with all these difficulties, and the great distance it is from the civilised world, the Onega Wood Company used to realise about 33 per cent. profit after having paid all expenses.'

The communication was accompanied with sections of trunks drawn to scale, from which it appears that from each trunk were obtained four boards of the following dimensions:—

Trunks in diameter, 7   7½   8   8¼   9   9½   10 vershi
Two inner boards, 9 × 3, 11 × 3, 11 × 3, 11 × 3, 11 × 3, 11 × 3, 11 × 3 inches
Two outer boards, 9-1¼, 11-1¼ 11-1½, 11-2¼, 11-2¼, 11-3, 11-3   ,,

The charge, royalty, or tax, as it is called in Russia, i. about 1s 4½d for every tree. Such is the charge throughou Russia, subject to modification by agreement or Imperia grant. And the tax, I was informed by the representative of the Company, was paid by them, and that on the same terms as any other purchaser. I was by another friend given to understand that while the Company was by their contract bound to fell or to pay for 60,000 trees, and in no one year to fell more than 200,000 trees, their annua operations were always much nearer to the latter number than to the former.

I had found in several of the annual reports of the Imperial Forest Administration an entry of 5000 rouble as paid by the Onega Company. This, I learned from the representative of the Company, was in earlier years a rent paid by the Company for the saw-mills, which were then the property of the Government, and was a charge distinct from that made for the wood; but the saw-mills were subsequently purchased from the Government by the Company, and the charge for rent ceased. But when the number of forest officials required to mark what trees should be felled was increased the Company was required again to pay 5000 roubles a year, to cover the expenditure, and a charge for extra watchmen in the Onega district. I had remarked in later Government reports an annual entry of charges for extra forest watchmen at Onega.

I was informed that oftener than once the Government had attempted to carry out the exploitation there; but it was always with a loss, and the existing arrangement was deemed more satisfactory.

By the Onega Wood Company sawn wood has been supplied for the market in France, and this may still be the case; but their trade is almost entirely with Britain.

When in St. Petersburg I learned that an enterprising and successful Russian timber merchant, either with his own capital or in combination with others, had completed arrangements for exploiting the forests in the far north upon a scale commensurate with those of the British Onega Company. The saw-mill was to be erected on the White Sea, and the necessary arrangements were being made. In these it was contemplated that thirty years would be required, and that thirty years would suffice, for their contemplated operations. Steam power was reckoned to be more economical than water power, for reasons which will immediately appear, and it was computed that the sawdust would supply the fuel required. As fuel this is preferred to the outside slabs of the timber, for this being generally damper than the sawdust obtained from the cutting-up of the timber, occasioned a waste of heat, and the draft of the chimney suffices to keep the sawdust in active combustion, though there may be a bed of it three feet thick under the boilers. The site of the saw-mill was determined by the facilities for getting the cut timber removed. The felled timber while uncut could be floated to the mill, cut timber must be otherwise transported, and there it could be shipped at once. There was water power to be had for nothing at various places nearer to the fellings; but then the transport of the cut material to the coast would cost money. The engineer laughed at the idea of portable or locomotive saw-mills, and said he had been employed in the manufacture of such, and had read flaming advertisements of their adaptation for employment in clearing out the timber in one district, and then being moved on to a second; and he showed the preposterousness of supposing that such a thing could be done there. I know forests in which it is otherwise; but I refer to the subject to show that facility of transport from the saw-mill is not of less importance than facility of transport to the mill, and may, as in this case, become a controlling element in deciding upon the operations to be undertaken.

The general arrangement was understood to be that so

much should be paid to Government for every thousand logs brought out of the forest, and something less for any felled, but rejected for defects subsequently discovered. A Government forest official superintended the delivery at the mill, and frequently another official in the same service marked what trees were to be felled. First, all trees above a specified girth at the upper end of a log cut 22 feet for 21 feet long, within a specified area of great extent, are first felled. Then this is done in one or more other areas. After this the same ground is gone over, felling trees of lesser bulk, and this process may be repeated, some of the areas being 100, 200, or 300 versts—70, 150, or 200 miles—from the saw-mill, and the fellings are so arranged by the forest official as to make the exploitation subservient to the preparation of the forests being managed in accordance with the most advanced forest science of the day, and the superseding of the method known as *Jardinage* by that known in France as that of *La Methode des compartiments*, the *Fachwerke method* of Cotta and Hartig.

The trees to be felled are first stripped of bark as a useless encumbrance, which is done roughly and speedily with the axe. In felling, the boll takes naturally a rounded form, but this is sawn across at the upper end, a foot being allowed in measuring to allow of the rounded end being cut off without encroaching on the required length of log, and the mark of the Company is then put on the upper end. The bark, branches, and other *débris* are left on the ground. All this is done in the winter season, because labour is then to be had. Arrangements are made with the officials of some village commune for bringing out the logs to the river side, also in the winter season. On the river they are made into rafts, and floated to the mill.

Many of the village communes have specified rights of felling trees in the forest for building purposes, for fuel, and even for sale. In a more advanced condition of forest economy, such rights acquired by usage or prescription are found to interfere seriously with the most advantageous management of forests, and they are being bought up in other countries on the continent of Europe. Here, in view of the future, they are not extended; and when an arrangement satisfactory to all parties can be made, they are being restricted or withdrawn; but meanwhile they are respected, and occasionally the officials of a commune come and offer to deliver say a thousand logs of specified dimensions at some specified point on a lake or river bank, or at the mill, on such terms as may be agreed upon. And such purchases are generally made in preference to felling on the contract with Government. Each raft consists in general of logs of uniform size, and thus the keeping of different sizes apart at the saw-mill is facilitated.

Thither they are transported in summer, and the cutting up is begun at once. Any taken out of the water and remaining not cut up at the end of the season, are afterwards returned to the water before being sawn, it being easier to saw them damp than dry. When they are cut up dry the teeth of the saw are often broken.

Of the forest operations of another company of timber merchants I received the following account :—' The members of this Company, the Messrs Thornton, had the land from the Government for the purpose of felling the timber, and were obliged by their contract to cut down so many thousand trees every year, paying so much per tree. It was for the contractors to judge what trees should be cut, and, so far as my informant remembered, without any Government supervision as to the trees.

'The timbers were dragged by horses to the nearest watercourse, but it did not pay to drag a tree nine versts to a stream. The trees were floated down to the mill, and, in this particular case, a waterfall was in the course, where many of the timbers were damaged, some even reduced to splinters.

'The sawm-ill was on an island, and in such a position that ships drawing about 25 feet could come alongside the mill. But in the case of timber that went by the River Onega, the mills were on the river, and the sawn timbers had to be taken in barges to the ships some distance out at sea. The mill was worked by steam-power —fuel being abundantly supplied by the saw-dust and wood unfit for shipping, with still enough remaining to have supplied firewood enough for the town of Kem, which was about six or seven miles distant.

'Tar is to some extent extracted from the *débris* of felled trees, but it scarcely pays for the manufacture. As a rule the *débris* is left to rot, as there no market for any products that might be obtained from it.

'By the terms of the contract the *débris* was to be removed from the ground, but the utter usefulness of it makes it more profitable to pay the Government's inspector, and keep on good terms with him, and then it rots where it is left, and he does not notice it.

'As a rule the people will not cut any trees more than 12 vershocks in diameter, so that such trees wherever they may be found are left standing. They seek for trees that will give 11 inch. planks, 3 inches thick, 21 feet long. Sometimes, I think frequently, they hew trees which will give two such lengths. One reason for leaving the thicker trees is that they are generally rotten at the heart, another is that frames are not made to saw larger timber. The wood was red pine. Their contract was only to fell timber wherever they found it profitable over a very considerable area. They had nothing to do with replanting, nor do they know whether any means were taken for that purpose. Nor did my informant know anything of how the trees stood on the ground, as he never saw them, nor did he see any on his journey to the place where the mill stood. It is a very profitable speculation. The cost of a tree at the mill might be about 1 or $1\frac{1}{2}$ roubles, but when sawn up into planks would be worth on the spot from 4 to 5 roubles.

'One reason why this affair did not turn out successful was this: It is usual in these contracts for clearing the ground for the Government to receive so much for each

21 feet length, but the friend who had led them into the business had so manipulated the people who drew up the contract that he paid so much *per root*. This made the superiors of the Department grumble; and they became suspicious, and put all possible difficulties in the way, and then circumstances arose which led the principal partners of the company to retire from the copartnery, and to leave the business in the hands of the manager, with what results I have not learned.

'On the River Mazeen one Rysanov has a contract for cutting the timber. He began five or six years back. At first there was a difficulty in getting ships to go there, but now they go readily. This is considered to be a wonderfully good business for Rysanov, as the wood is of an uncommon kind for that part.'

By another of my informants, not less conversant with the work, I was told that in some cases the whole of the wood belonging to several wood-cutters or dealers is brought to one place on the banks of a stream, some as logs, some in billets for furnace fires, $2\frac{1}{4}$ arschens in length, some in billets for household use, or not marked as belonging to different proprietors, but all carefully measured, and the measurements of wood belonging to each carefully noted, and all is floated away together to be re-collected by a weir and proper appliances at a lower level, where it is again divided in the proportions noted, and any deficiency is borne by the whole in the same proportions. But the large timber is generally floated on the lower rivers in rafts. Much of the wood is floated, and in some parts, but not everywhere, a charge is made by Government for license to float the timber.

By another gentleman, an engineer, I was informed that he had not seen the cutting of trees for timber, but he had seen a good deal of various saw-mills at the mouths of rivers flowing into the Lake Ladoga. The timber, roughly hewn, is floated down these rivers towards the lake. At rapids there are great works constructed for the shooting of the timber, and lower falls near the mouths of the rivers are utilised as a moving power for saw-mills erected there, where the rough timber is cut up chiefly into three and four inch planks, in three, four, and five fathom lengths, seven feet being the length of a fathom. It is there laden on Ladoga vessels, fitted with sails, and is then conveyed by the lake or by canal to the Neva, and by that river to St. Petersburg or Cronstadt.

Wood and timber cutting, he added, is carried on wherever there are means of water transport to the basin of the Neva. The firewood thus obtained is conveyed to the city in barges, which are made to be there broken up and sold as coarse planking, and the timber is conveyed in rafts. A good deal of Baltic timber is brought to St. Petersburg from the lower ports, and some wood comes from the Gulf of Finland, but the bulk of what is there sold is brought down the Neva.

The same gentleman wrote to me in another letter :— 'On the north shore of the Lake Ladoga I saw an enormous saw-mill, with timber enough in its structure to have gone a great way in many different works in Britain. A big, jolly-looking Russian peasant, after showing it all, said, his face radiant with exulting satisfaction, "*I made it all myself.*"'

Mr Judræ says :—'In connection with this subject, the following statement may show approximately what are the proceeds of the sawing of timber. From four logs are produced three dozen of boards of different measurements. Four logs, according to the present charge, cost 1·80 rs.; the transport to the river and flotage, sawing, shipment, and freight to Cronstadt of these cost 10 rs.; so the total cost is 11·80 rs., and the three dozen boards at Cronstadt are worth 18 rs. But the calculation, it must be borne in mind, is only an estimate approximately correct.

M. Werekha, in the work already cited, says :—
'Most of the sawing is done by hand. The greater part

of the machines are hydraulic mills, but of late years the number wrought by steam-power has considerably increased. There are reckoned to be in Russia about a hundred great saw-mills. We know of thirty driven by steam, and of this number there are six in the Government of Archangel which are occupied solely in sawing wood of wild pine for the commerce with England. In the number of saw-mills driven by water there are some which work only for the supply of local requirements, and in small quantities; but there are thirty steam mills, and fifty water-mills, which work chiefly for the foreign trade. In the whole of these mills they cut up annually at least two millions of trees. The cutting up of trees into planks, for local requirements, is done often in the place itself, or where the exploitation of the forest takes place. The Governments of Kostroma, of Kazan, of Viatka, and the southern part of that of Vologda, furnish planks to the less wooded countries situated on the lower course of the Volga and countries on the Don. The Governments of St. Petersburg, of Olonetz, and a portion of that of Novogorod, export planks by the port of St. Petersburg, and furnish to this city logs to be sawn. From the northern part of the Government of Vologda logs, round or square, are exported by the ports of the White Sea.'

In the *Ustaff Laesnoi*, or Forest Code, are laid down regulations for the administration of Crown forests. In this Code No. 677 refers to saw-mills, private property, supplied with timber from Government forests on payment of rent; and in the appendix it is stated that there were, in 1857, such mills situated in the following places in the Government of Olonetz: at Gorsk, Tulodgsk, Vedlozersk, Kotkezersk, Zadnenikiforsk, in Petrozavodsk, at Syvatozersk, Svyamozersk, Lindozersk, and Kijsk, in Novaenetzkom, and at some other places, at some of which a rent of 300 roubles for each saw is paid per month; at the others, 250 roubles for each saw is paid.

In Rule 6 it is stated: Experience has shown that in

first-class mills, two frames, well-managed, and in full work, will saw from 3000 to 3480 trees per month; and in the second class from 2000 to 2800 per month And in view of this, the Forest Department of the Ministry of Imperial Domains requires the holders of the saw-mills to state beforehand how many frames they design to use, and a license is given in accordance with his notice to fell, but to fell only so much timber in advance as he can cut up in a year or in a year and a half. And should he be found to have felled more than his license allows, on proof of this after due investigation, he shall be regarded as an intentional depredator. Of trees in which the branches are high, trunks twenty-two feet long, and six vershocks in diameter, is the normal measurement. Of trees clothed with branches to near the ground, the normal measurement is a trunk fourteen feet long, and six and a-half vershocks in diameter. Sixteen vershocks is equal to one arshin, or twenty-eight inches.

The millowner is free to use timber for repairs of the mill, but not for building a new mill.

In the list of changes made previous to 1863, appended to the Code, it is stated that in 1860 an order was issued that the wood should be disposed of to the owners of saw-mills by auction, and that this arrangement, which was experimental, should be carried out for a period of six years, but this order was rescinded on 1st May 1861, and was not again renewed.

In the list of changes made previous to 1868, it was stated that in 1866 certain matters pertaining to the Imperial Domains were, excepting in the Baltic Provinces, transferred to the supervision of special administrations in the several Governments, and amongst these was jurisdiction in regard to this statute.

In the Code No. 773 prescribes permit tickets which wood merchants must obtain. In the appendix are regulations laid down in regard to permit tickets required for the transport of wood by rivers, and in regard to the permit tickets required for the transport of wood by land.

No. 1269 refers to the practice of wood-cutters leaving a high stump standing, and enjoins upon the official in command to see to the conversion of cleared forest into arable land.

In the appendix regulations are laid down for the preservation of Government forests from destruction in connection with the working of the gold mines in the Government of Vologod, enjoining that no more trees than necessary shall be felled, and in mining to conduct operations as to allow the trees to grow on the surface. In the list of changes made previous to 1871, it is stated that this statute had been changed.

In the appendix to No. 1471 is given a tabular statement of the fines to be exacted for the felling of timber without permission, graduated according to the length of the trunk in sajeens, or fathoms, and the diameter in vershocks, and ranging from 1 rouble, 60 kopecs for a tree three fathoms long, and four vershocks in diameter, to 113 roubles, 80 kopecs for a trunk ten fathoms long and nineteen vershocks in diameter.

The *sajeen*, or fathom, may be reckoned seven feet; the *arsheen*, 28 inches; and the *vershock*, 1¾ inches. The *desatin*, a land measurement, is equal to about 1¼ acres; and a *verst* is equal to two-thirds of a mile.

CHAPTER V.

EXPORTS BY ARCHANGEL AND THE WHITE SEA.

WHILE the timber cut in these regions, and in others adjacent to them, finds its way to St. Petersburg, and of this some is exported thence, most is exported direct from the ports in the White Sea. One of the most important export ports is Archangel, situated on the Dwina, and the gulf known as the Gulf of Archangel, but named by the Russians from the river, the Dvines Kaia Gulf. Of this great outlet for forest produce Hepworth Dixon supplies, in his account of his approach to Russia, and entrance by that northern haven, the following account :—

'At Cape Tutsi we pass from the narrow straits dividing the Lapp country from the Samoyed country into this northern gulf. About twice the size of Lake Superior in the United States, this Frozen Sea has something of the shape of Como; one narrow northern bay, extending to the town of Kandalax, in Russian Lapland, with two southern bays, divided from each other by a broad, sandy peninsula, the home of a few villagers employed in snaring cod and hunting seal. These southern bays are known, from the rivers which fall into them, as Onega Bay and Dvina Bay. At the mouths of these rivers stand the two trading ports of Onega and Archangel.

'The open part of this inland gulf is deep—from 60 to 80 fathoms; and in one place of the entrance into Kandalax Bay the line goes down to no less than 160 fathoms. Yet the shore is neither steep nor high. The Gulf of Onega is rich in rocks and islets, many of them only banks of sand and mud, washed out into the sea from the uplands of Kargopol; but in the wild entrance of Onega Bay, between Orlof Point and the town of Kem, stands out a notable group of islets—Solovetsk, Anzersk, Moksalma, Zaet, and others: islets which play a singular part in the history of Russia, and connect themselves with curious legends of the Imperial Court.

'In Solovetsk, the largest of this group of islets, stands the famous convent of that name; the house of Saints Savatie and Zosima; the refuge of St. Philip; the shrine to which emperors and peasants go on pilgrimage. . .

'By the Maimax arm we steam through the Delta for some twenty miles, past low green banks and isles, bright with grass and scrub. Beyond them, on the mainland, lies a fringe of pines going back into space as far as the eye can pierce. The low island lying on your right, as you scrape the bar, is called St. Nicholas, after that sturdy priest, who is said to have smitten the heretic Arius on his cheek. . . . .

'On passing into the Maimax arm, your eyes—long dimmed by the sight of sombre rock, dark cloud, and sullen surf—are charmed by soft green grass and scrub; but the sight goes vainly out through reeds and copse, in search of some cheery note of house and farm. One log hut you pass, and only one. Two men are standing near the bank, in a little clearing of the wood; a lad is rolling in a frail canoe, which the wash of your steamer lifts and laves; but no one lodges in the shed. The men and boy have come from a village some miles away. Dropping down the river in their boat to cut down grass for their cows, and gather up fuel for their winter fires, they will jump into their canoe at vespers, and hie them home.

'On the banks of older channels the villages are thick; slight groups of sheds and churches, with a cloister here and there, and a scatter of windmills whirling against the sky. . . . .

'On all these banks you notice a forest of memorial crosses. When a sailor meets with bad weather he goes on shore and sets up a cross. At the foot of this symbol he kneels in prayer, and when a fair wind rises he leaves his offering on the lonely coast. When the peril is sharp, the whole ship's crew will land, cut down and carve tall trees, and set up a memorial with names and dates. All round the margins of the Frozen Sea these pious witnesses abound; and they are most of all numerous on the rocks and banks of the Holy Isles. Each cross erected is the record of a storm. . . . .

'Climbing up the river you come upon fleets of rafts and praams, on which you may observe some part of the native life. The rafts are floats of timber—pine logs, lashed together with twigs of willow, capped with a tent of planks in which the owner sleeps, while his woodmen lie about in the open air when they are not paddling the raft and guiding it down the stream. These rafts come down the Dvina and its feeders for a thousand miles. Cut in the great forests of Vologda and Nijni Konetz, the pines are dragged to the water-side, and knitted by rude hands into these broad, floating masses. At the towns more sturdy helpers can be hired for nothing; many of the poor peasants being anxious to get down the river on their way to the shrines of Solovetsk For a passage on the raft these pilgrims take a turn at the oar, and help the owners to guide her through the shoals.

'In the praams the life is a little less bleak and rough than it is on board the rafts. In form the praam is like the toy called a "Noah's Ark;" a huge hull of coarse pine logs, rivetted and clamped with iron, covered by a peaked planked roof. A big one will cost from 600 to 700 roubles (the rouble may be reckoned for the moment as half-a-crown), and will carry from 600 to 800 tons of oats and rye. A small section of the praam is boarded off to be used as a room. Some bits of pine are shaped into a stool, a table, and a shelf. From the roof-beam swings an iron pot, in which the boatmen cook their food while they are out on the open stream, and at other times—that is to say, when they are lying in port—no fire is allowed on board, not even a pipe is lighted, and the watermen's victuals must be cooked on shore. Four or five logs lashed together serve them for a launch, by means of which they can easily paddle to the bank.

'Like the rafts, the praams take on board a great many pilgrims from the upper country; giving them a free passage down, with a supply of tea and black bread as rations, in return for their labour at the paddle and the oar. Not much labour is required, for the praam floats down with the stream. Arrived at Solaubola, she empties her cargo of oats into the foreign ships (most of them bound for the Forth, the Tyne, and the Thames); and then she is moored to the bank, cut up, and sold. Some of her logs may be used again for building sheds, the rest is of little use except for the kitchen and the stove.'

'Like all great rivers,' says Mr Dixon, 'the Dwina has thrown up a delta of isles and islets near her mouth, through which she pours her flood into the sea by a dozen arms. None of these dozen arms can now be laid down as her main entrance; for the river is more capricious than the sea; so that a skipper who leaves her by one outlet in August, may have to enter by another when he comes back to her in June.'

Interesting, amusing, and saddening narratives are given relative to the arm by which he entered the river, some of them illustrative of the difficulties of dealing with provincial authorities in Russia, both in relation to trade and to matters of perhaps more importance; and he goes on to say:—

'In catching a first glimpse of the city of Archangel, you are struck by the forest of domes and spires; the domes all colour and the spires all gold. . . . On feeling for the river-side a captain finds no quay, no dock, no landing-pier, no stair. He brings to as he can, and drags his boat into position with a pole. . . . .

Archangel is not a port and city in the sense in which Hamburg and Hull are ports and cities, with clusters of docks and sheds, with shops and waggons and a busy private trade. Archangel is a camp of shanties, heaped around groups of belfries, cupolas, and domes. Imagine a vast green marsh along the bank of a broad brown river, with mounds of clay cropping here and there out of the peat and bog; put buildings on these mounds of clay; adorn the buildings with frescoes, crown them with cupolas and crosses; fill in the space between church and convent, and convent and church, with piles and planks, so as to make ground for gardens, streets, and yards; cut two wide lanes from the church called "Smith's Wife," to the monastery of St. Michael, three or four miles in length; connect these lanes and the stream by a dozen clearings; paint the walls of church and convent white, the domes green and blue; surround the log houses with open gardens; stick a geranium, a fuchsia, and an oleander into every window;. leave the grass growing everywhere in street and clearing —and you have Archangel.'

In a work entitled, *The Land of the North Wind; or, Travels among the Laplanders and the Samoyeds*, by Edward Rae, F.R.G.S., the following account is given of Rusanovna, another port of export which he visited in 1874:—

'Three years ago there was nothing whatever here but a steep muddy bank, crowned with firs of the virgin forest. Mr Rusanoff, a man of ability and initiative, coming here, was struck by the advantages the spot possessed for the establishment of a timber port. After long and patient investigations in the district he took from the Russian Government, I am informed, a concession of eight million *desatin* of forest land lying on and about the great river and its branches. This is equivalent to fourteen millions of acres, which is considerable. The area of France, inclusive of the two provinces temporarily occupied by the Germans, is about two hundred thousand square miles, and we should like to make a comparison; but as neither my companion nor I know how many acres there are in a square mile, and don't mean to learn superficial measurement until the métric system is introduced into our puzzleheaded fossil old country, we must leave the question alone.

Rusanova, with its capabilities, will develop the resources of this district.

'Mr Rusanoff has two tug steamers and a number of barges: the steam saw-mills are capable of cutting sixty thousand trees, representing a quarter of a million of planks, in a year. In addition to the church Mr Rusanoff erected a schoolroom, an important store for provisions and other necessaries, large house accommodation, and then commenced his business. The trees, hewn in the primæval forests around, are lashed into rafts of perhaps two hundred each, and floated down to the mouths of the rivers, where the steamers go to take them in tow. Arrived at Rusanova, they pass through the saw-mills, and are ready for shipment abroad. Once commenced the operations soon began to grow. In the first year several ships came for timber; last year sixteen came; this year, the third, twenty-two large ships and nine smaller vessels are to come; next year Mr Rusanoff's business engagements will require fifty ships.

'Three years ago the value of labour here was fifteen kopecks, or fivepence a day; now it is worth a rouble, or two shillings and ninepence a day. The port is an excellent one. At low tide there are nineteen feet of water in the channel abreast of the quay, at high water from thirty-eight to forty-four feet, according to the height of the tide. There is no bar, and beyond Masslynnoi Nôs, the pilot station and beacon seven miles away, is the deep sea. Mr Rusanoff means to construct this winter a tall lighthouse and life-boat station upon Masslynnoi Point, to replace the beacon, and perfect the means of access to the port. The approach of ships is signalled from the beacon, and the steamers are always available for towing ships at a moderate cost. The daylight during the open navigation is practically constant, and the saw-mills and steamers work night and day. The harbour was open last year considerably earlier than Archangel, ships coming here when the other port was closed. The difficult and often tedious voyage down the White Sea, and the miserable approaches to Archangel are avoided, and the voyage to England is two hundred miles shorter. The average of voyages of English ships coming here for timber has been twenty-eight days.'

CHAPTER VI.

FOREST INDUSTRIES.

Section A.—Forest Exploitation and Clearing of Forest Lands.

Mr Judræ, in the accounts of his journey of inspection in the Government of Olonetz, has made mention of complaints of the timber trade being unremunerative. He writes:—' Speaking generally, the first acquaintance of one with the country leads to the conclusion that the Government of Olonetz is as poor in works employing human industrial labour as it is rich in natural productions, amongst which the first place must be assigned to those of the forests. In Petrozavodsk I was enabled to collect from records by officials who had formerly the management of the forests, and of all matters relating to the country, information of which the following is a summary.

'The Government of Olonetz lies between 60° 21' and 65° 16' N. Lat., and 47° 21' and 59° 36' E. from the meridian of Faro, corresponding to about 30° and 40° E. of Greenwich. It has an area of 2,785 square geographical miles, or 14,026,320 *desatins*. Of this area forests cover approximately ten millions of *desatins*, or five-sevenths of the whole. After deducting 257,000 *desatins* of arable land, and 88 *desatins* of pasture land, the rest is composed of rivers, lakes, swamps, and other unproductive places. The whole population, including both sexes, amounts to 301,290 ; consequently there is for each man 47 *desatins* of surface, consisting of—

> 1·14 *desatins* of arable and pasture land ;
> 35·19 of forests ; and
> 10·67 of lake and river.

'This proportion of the population to the area is indicative of the poverty of the territory. The forests belonging to the Imperial Domaines measure 8,774,419 *desatins* and 1,048 square fathoms, or about 1,740 square geographical miles, which amounts to about two-thirds of the whole area of the Government. In 1865 the revenue derived directly from the forests amounted to 327,993 roubles,* and by extra fellings 9,607 roubles 90 kopecs; in all, 337,540·90 roubles. According to calculation each *desatin* on an average yielded a revenue of 3·84 roubles. In subsequent years the revenue was considerably diminished in consequence of the saw-mills not working.

'In so far as forests are concerned, the importance of the Government of Olonetz is seen more in view of the future than in relation to the present. Having several navigable outlets, it may be considered a reserve of forests available not for Russia only, but for Europe.

'Looking into the accounts of revenue derived from these forests, we find that almost 45 per cent. of the revenue is the proceeds from the sale of timber taken to the sawmills. In 1865 there were sold to seven of these 237,783 logs for the sum of 98,359 roubles 59½ kopecs, and to the English Onega Company, having its *fabrique* on the River Onega in the Government of Archangel—but preparing at the present time forest material in the district of Kargopol in the Government of Olonetz,—logs amounting in value to 52,585 roubles; in all, 150,944 roubles 59½ kopecs.

'From what has been said it follows that the saw-mills, which are the principal purchasers, are indispensable for the sale of timber; and that but for these there would be but a small sale of timber, more particularly in the northern parts of the Government.'

This is irrespective of the expense of transport in regard

---

* The standard equivalent of the rouble is 3s 4d. It is generally, in accordance with the rate of exchange, 2s 6d. When I was in Russia last year it was 2s, and at one time during the war it was 1s 10d. The rouble is equal to 100 kopecs.—J. C. B.

to which he makes some important statements, some of which have been cited. Beyond all question the timber trade is at present the most important throughout the whole region. With increased facilities for transport there may be drawn hence a large supply of fuel, so soon as the rise in price occasioned by the diminishing supply in the central and southern Governments of Russia may make it remunerative to send firewood to the capital.

In regard to terms on which the permission to fell timber is given, information is embodied in the accounts given by me of the operations of different companies engaged in the trade. On this point the following more general information was obtained by me. Where wood is in demand as an article of commerce, whether as timber or as firewood, standing forests are sometimes, both by Government and by private proprietors, disposed of at so much per tree felled, or so much per fathom of firewood obtained. Sometimes the charge is made for permission to fell for a specified time, embracing, it may be, several years. In the arrangements made relative to felling, sometimes the trees which it is permitted to fell are marked by a representative of the proprietor, but more frequently the licence holder is allowed to fell what and where he pleases. In the one case, and in the other sometimes, only such trees as may be preferred are felled in accordance with the method of exploitation known as *Jardinage;* in other cases a measured area is cleared entirely.

In the vicinity of the Urals, in the Government of Perm, forests are exploited in accordance with the method known in forest science as *à tire et aire:* in successive decades, successive portions are cleared with the exception of *balliveaux*, or reserved seed bearing trees of mature age, and it may be saplings, the forest being divided into such number of sections that by the time the woodman may have gone over the whole the first cleared section will be again ready for the axe.

This is not practised here, but in some places there is adopted systematically what is an approximation to it, in accordance with a practice followed there and elsewhere in the distribution of allotments of communal arable ground. In this the ground is divided into long narrow strips or lines from three to six fathoms broad, and from one hundred to five hundred long, which strips are again sub-divided into lots of equal size for allotment to the members of the community. Something similar may be seen in the vicinity of Berwick-upon-Tweed, where lands granted to the freemen by James I. of England are allotted periodically to individual members of the community.

In other cases in Russia, it may be for convenience in the tillage of lands allowed to lie many years fallow, these are cultivated in long strips, which, varying in colour with the crop grown, or the years which have elapsed since they were tilled, present to the eye of the passing traveller what may suggest the idea of a corduroy of variously-coloured ridges. In felling or clearing forests something similar is done—long straight strips being cleared, with strips of forest between them, sheltering from destructive storms the crops which are raised.

Sometimes the permission to fell the trees on ground to be cleared is disposed of by auction, sometimes it is otherwise. In sales by auction it is assumed—it may be the result of what is known in forest science as *taxation*, survey and measurement—that the whole area, or so many decatines, is forest, and that each decatin contained so many trees, or so many cubic fathoms of wood ; and, according to what may be the terms of sale, the offer of the buyer may, or may not be subject to deduction, either for deficiency in extent, or deficiency in number or cubic contents of trees ; but the buyer has the benefit of any excess over what had been assumed. But the whole of these conditions and details pertain more to private than to State forests, and the mention of them is leading us away from the regions which are here more particularly under consideration.

Of several forest officials I enquired whether the annual production of wood in the district equalled, exceeded, or was less than the consumption by felling and fire and waste. The general reply was that there had not yet been so complete a taxation or estimate of the cubic contents of existing trees, of their numbers, and of their annual cubic increase of growth, as would warrant a definite statement; but the opinion of two was that the production was equal to the consumption and destruction; the opinion of a third was that it was not, and consequently that the mass of wood in the forests was being annually diminished.

Section B.—Tar, Turpentine, and Vinegar Manufacture.

Besides the felling of timber for transport to a distance, there are other forest industries carried on in this region. There is wood felled for use as fuel, and for the manufacture of vinegar, tar, and other products.

Mr Judræ, in his account of a journey from Vosnecenya to Petrozavodsk, says:—'The first thing which interested me was the forest-product manufactory of Mr Baelaeff, well known in all these northern parts. It is situated about seven versts from Vosnesenya.

'The lovely view presented by the Fabrique and buildings around leads me to conclude that it must be a profitable property, yielding a considerable revenue. It is built in a situation very convenient for the sale of the products; near to Vosnesenya, where there is a great consumption of tar in caulking vessels. Hitherto there could only be obtained black burnt tar, which is not quite suitable for the purpose, and the demand for it was not great; but now they are constructing new brick furnaces for the production of what is called red tar, from the sale of which they will, without doubt, obtain considerable profits.

'There is not a scientific or special manufacturer employed, but the works are under the management of an able workman; by this arrangement it is supposed a great saving is effected. The Fabrique contains at present several furnaces, by which are obtained tar, turpentine, and other products from pine wood. Besides these, there are furnaces for rectifying turpentine and for making pyrolignous acid. The latter product is obtained from birch wood by a process of dry distillation. The greater part of this product is taken to St. Petersburg, but the greater part of the red tar commands a sale in the locality.

'The following are details relating to the manufacture of such articles obtained on the spot. From a cubic fathom of wood are obtained 25 poods of black tar, equal to 900 lbs. English, and two poods or 72 lbs. of turpentine; and in the manufacture they consume half a cubic foot of firewood.

'From a square fathom of birch they obtain 250 [?] poods of pyrolignous acid.

'How far these figures indicate the reasonableness and propriety of the measures adopted may be determined by a comparison of them with results obtained by scientific operations, and with the returns made by other works of the same kind elsewhere. The proprietor was desirous of impressing on me that the establishment is not remunerative, and hardly returns the working expenditure. The quantity of acid manufactured is some hundred tons more than suffices to meet the demand for it, and the turpentine is scarcely equal in quality to what is required in the market, and thus he accounted for the unremunerative character of the works.'

I have visited this work, and the result of subsequent years' experience seems to have proved that Monsieur B., the proprietor, was correct in his views. The place was to some extent in ruins; there were piles of pine-tree roots for the production of tar, and piles of birchwood for the production of vinegar, but no work had been done there for years. A solitary workman lived there in charge of the place. The apparatus for the manufacture of tar had been removed, and the supports for the retorts were in ruins. This may also be said of the vinegar works, but there was a new retort ready for erection, and it was intimated that the proprietor had some intentions of resuming this industry. Tar is manufactured extensively in the Government of Archangel. The operation is of the simplest character.

Spirits of turpentine are also manufactured there, and this may also be reckoned among the small industries of the peasants living in forest districts. The following account of how this was done sixty years ago may be considered antiquated, but amongst a population such as they are changes in rural industries are not frequent, or speedily and extensively effected. It is extracted from a paper published in the Transactions of the Highland Society of Scotland for 1820, entitled 'An Account of the Manufacture of Turpentine from the *Pinus Sylvestris*, as practised by the Native Peasantry of the Interior of the Russian Empire.' By William Howison, M.D.

'The second day after my arrival,' writes Dr Howison, 'I made an excursion in the neighbourhood of the mansion-house, during the course of which I arrived at a wretched building, situated upon the margin of the forest, at the door of which two Russian boors were busily employed with their hatchets in cutting into small chips the stumps and dried roots of fir trees, which had been previously dug from the earth, and were lying collected together upon the surface of the snow. Upon going into the interior of the wooden shed or building, there was a fine clear fire burning, and two old boors distilling turpentine from the chips of fir wood broken down, as already noticed, by their companions. In the centre of the apartment there was a brick furnace, with a clear fire burning in it, and a large iron boiler built in above it. The boiler was completely filled with the cut chips of wood, and a quantity of water; the flame of the fire reverberating upon its under surface. From the top of the boiler, which was accurately and neatly covered up with a close lid, a spiral iron tube passed out, and entered a large wooden vessel placed within a short distance from it, which originally had been completely filled with snow and ice, but which, by this time, were almost entirely converted into warm water, by condensing the heated tar and vapour which passed from the boiler. Within the vat this spiral tube formed a tortuous worm; and again passing out at the opposite bottom of the vessel, to the end of it a long glass bottle was luted, which received the turpentine as it dropped from the tube. One side of the house was filled with the recent cut chips of the fir wood, which had not as yet been put into the boiler; whilst the other side contained those which had come out from it, from which the turpentine had been extracted, and which were now used as fuel to supply the fire.

'A little after my arrival the distillation was completed, and the boors removed the bottle, which was rudely luted by means of clay, from the tube. Upon examining its contents I found that the under half of it contained water, whilst the upper one contained the empyreumatic oil of turpentine, which, from its less specific gravity, naturally rises towards the surface. In order to separate it from the water, these Russian boors took a very simple method, and, at the same time, one very characteristic of a barbarous people. The bottle, which was of coarse green glass, had a very minute hole bored in the bottom of it, which was stopt up with a small wooden plug. They removed this plug, and allowed the water gradually to escape, until the turpentine made its appearance at the hole, when they replaced the pin, and poured the turpentine into another bottle for preservation; which constituted the whole process.

'Upon requesting to see the quantity of turpentine which they had made in the course of the day, the old Russian brought from the corner of the house a bottle, which might contain from four to five pounds, if my memory does not mislead me; and this, as already mentioned, was entirely procured from the stump and roots which remained after the trunk was cut down, and which could be applied to no other use.

'Distilling houses, similar to that now described, are to be met with upon the estates of the different noblemen, or landed proprietors, in the northern parts of the Russian empire. Consequently, an immense quantity of turpentine must be procured in this way during the course of the year, both for public and private consumption. It produces a great advantage also in affording in-door work for the boor during the severity of a long and dismal winter.'

SECTION C.—HOUSE BUILDING AND CARPENTRY.

Throughout the district, as is generally the case—I had almost said throughout the whole of Russia—the houses are built of logs laid one upon another, and caulked with moss, those of adjacent sides crossing each other a little way from the corner; and wood is the only fuel used. I have visited at houses elegantly furnished, which must have been done at great expense, and where the dress, accomplishments, rank, and bearing of the inmates and their visitors were such as one might expect to meet with only in the more fashionable resorts of Central Europe, but where the houses were only such as I have referred to —elegant and somewhat imposing in their external aspect, for which the mode of structure offers facilities; but internally even the public rooms had walls and partitions of slightly hewn logs, without covering of paint, tapestry, or paper.

In these the furniture was made to some extent of imported woods—rosewood and mahogany—but largely of the forest produce of the locality.

In Vologda, and in all the forest lands of the north-eastern districts, all the world is *plotnik*—a carpenter, and these carpenters, who work in wood in every possible fashion, manufacture the most delicate articles as well as the rudest, with their hatchets alone, and hardly ever using their saws. Their ability and cunning workmanship, remarks Wahl, are qualities not to be met with in any foreign country, and must excite the admiration of all beholders.

PART III.

PHYSICAL GEOGRAPHY.

In geography there are discussed several matters pertaining to different categories: mathematical, physical, and political. To the first-mentioned category belong questions relative to the earth as part of the solar system. To the last-mentioned belong questions relative to the human population in their relations as subjects of different kingdoms, or citizens of different states. Under the head of physical geography are discussed questions relative to the geological formation; the contour of divisions of a country, more or less extensive; the phenomena of tides and currents; modifications in the atmosphere with regard to weight, temperature, humidity, and motion; and, in connection with this, the flora, fauna, and ethnographical relations of inhabitants of different countries.

In the following statement I shall follow generally the division thus indicated; but some of the matters are so correlated that notices of matters pertaining, strictly speaking, to one, may find its place amongst details given of matters pertaining to another, where I find this convenient and justifiable; and what relates to the ethnography of the region I may afterwards bring under consideration in a separate work.

CHAPTER I.

CONTOUR AND GENERAL APPEARANCE OF THE COUNTRY.

FROM the account I have given of my voyage from St. Petersburg to Petrozavodsk, and of my trip thence to the Falls of Keewash, and from the narratives given by Messrs Judræ and Hepworth Dixon of their journeys through the forests which they traversed, there may be gathered a pretty correct idea of the contour and general appearance of the western portion of the forest zone of Northern Russia, and more especially of those of the Government of Olonetz.

To the east of Olonetz is the Government of Vologda, extending thence to the Ural Mountains, between the Governments of Archangel on the north, and those of Perm, Viatka, Kostroma, and Yaroslaf on the south, with an area of 337,111 square versts, or 150,000 square miles.

The surface is generally flat. Mountains are rare, but a succession of hill and dale is very common; and in many places these inequalities produce scenery which is not deficient in beauty. Nearly all the rivers belong to the bason of the Arctic Ocean. The principal are the Dwina, the Sukhona, the Louza, Vega, Vitchegda, Mezen, Pisega, and Petchora. The Government takes its name from the river Vologda, which, taking its rise from a marsh, flows into the Sukhona on the right bank, after a course of 90 miles. The Louza, rising 90 miles east of Nikolsk, and flowing north-east and west, passes Lalsk, and joins the Joug, 18 miles south-east of Veliki-Oustioug.

A small portion of the Government in the south is drained by affluents of the Volga.

In the northern parts of the Government the trees lose their leaves in August, and the rivers are frozen over from the end of October to the middle of April. In the south there are large tracts occupied by forests, lakes, and morasses. Agriculture is followed to a certain extent, but the severity and changeable state of the weather render it precarious, though a considerable quantity of wheat and barley are grown. The produce of the pasture grounds, of the chase, and of fishing, tend to compensate for this, and the woods supply potash, tar, and other materials for export as well as domestic use.

Of the general appearance of lands in the extreme north, some idea may be formed from the details given by M. Guillemard, and Mr Hepworth Dixon, cited in a preceding chapter. To the north of the Government of Olonetz is the Government of Archangel, stretching from Finland on the west to the Ural Mountains and the Government of Tobolsk in Siberia on the east, comprising thus the whole northern part of Russia in Europe, and including the island of Nova Zembla. Its northern continental shores are washed by the Arctic Ocean and the White Sea; and for a considerable distance from the coast they present a desolate and sterile appearance, with few signs of vegetation. The surface of the remainder is in general a continuous flat, covered either with sandy and mossy wastes, or pine and alder forests. The area, inclusive of that of the islands, which has been spoken of as about a fourth of the whole, has been estimated by Möller at 15,215 German, or 342,337 English square miles; by Kœppen it has been estimated at 15,519 German square miles.

The river Onega is a large river rising in Lake Latcha, to the east of Lake Onega, and flowing thence north-west it falls into a gulf, in the White Sea, to which it gives its name, as it does also to the town at its mouth, about 80 miles S.S.W. of Archangel. Its principal affluents are the Voloshka and Mokha on the right bank, and the Kena on the left.

There are several lakes in the Government of Onega. Amongst others the Imandra, in the district of Kola, which is 60 miles in length from south to north, and about 15 miles in breadth, and it discharges its waters into the White Sea,—the Taposero, the Angosero, and the Koutno.

The principal rivers are the Onega, the Dwina and its affluents, the Petchora, and the Mezen. The Dwina is formed by the junction of the Sukhona and Joug, navigable rivers coming, the former from the Scandinavian Alps, and the latter from the Ural Mountains. The confluence occurs a little below Veliki-Oustioug, in the Government of Vologda. They are subsequently joined by the Solvytchegodsk, on which is situated a town of the same name. And the united waters, in crossing the Government of Archangel, pass Kohlmogori and Archangel, and flow by several mouths into the gulf of the White Sea, which bears the name of this town. The total course of this river, one of the largest in Russia, is 420 miles, and its greatest breadth is five miles. Its depth is considerable, but its navigation is impeded by beds of mud which bar its embouchures, and by the number of islands with which, throughout the greater extent of its course, its channel is obstructed. The tides extend to a distance of 30 miles above Archangel. The Dwina was for a long time the only outlet for the productions of European Russia. The country through which it passes is low and level, and is to a great extent laid under water by its inundations in the spring. Of the numerous affluents the principal on the right bank are: the Vytchegda, the Ourtiouga, the two Toïma, the Vaengha, Pinghicha, Poukchenga, Pinega, and Lodma; and on the left the Oustioumej, the Kodima, Vaga, Emtsa, and Laïa.

The Vytchegda issues from a marsh in the district of Oust-Sisolsk, in the east part of the Government of Vologda, and after a total course of 450 miles, flows into the Dwina a little below Solvytchegodsk. Its principal affluents are the Yulva and Yarenga on the right bank, and the Sisola on the left. It is at all times navigable.

The Pinega, which gives its name to a town situated about 78 miles east by south of Archangel, flows into the Dwina after a separate course of 250 miles.

Connections have been formed between the Dwina and the Volga by means of canals, one of which joins the Keltma, one of the head streams of the Vytchegda, with the Kama; and the other, known as the Lubiuski Canal, unites the Sukhoma with the Neva by means of the Cheksna.

The Petchora has its source in the Government of Perm, on the west side of the Ural Mountains, and crossing the Government enters that of Vologda, and that of Archangel, and, after a tortuous course, flows into the Arctic Ocean by numerous mouths after a course of upwards of 900 miles. The country through which it flows is low, covered with wood, and nearly uninhabited. Its principal affluents are, on the right, Ilicha and Oussa; on the left, the Ijma and Tsylma. The Ijma rises in the Government of Vologda, enters the Government of Archangel in the district of Mezen, and flows into the Petchora on the left bank, after a course of 240 miles.

The Mezen has its source in the district of that name; it afterwards enters the Government of Vologda, but returning and passing the town to which it owes its name, after a total course of about 480 miles, it flows into the Gulf of Mezen, an arm of the White Sea, 75 miles wide at its mouth, and indenting the land to a depth of 60 miles. The principal affluents of the Mezen are the Piema and Peza on the right, and on the left the Vachka. The Peza has its source in a marshy locality, and enters the Mezen at Jerd, 36 miles above the town of Mezen, after a course of above 180 miles. The Vachka takes its rise in the Government of Vologda, in the district and to the northwest of Jarensk, and joins the Mezen near Oust-Vachka after a course of 225 miles.

The district of Mezen occupies the eastern part of the Government. It is 600 miles in length from west to east, and upwards of 300 miles in breadth on the mainland, that is exclusive of Nova Zembla, which depends upon it, and of the islands Kalgoner and Vaïgatch. It has a level surface traversed by the Petchora and Mezen, and contains numerous marshes, and has on some parts good soil and abundant pasturage, but the severity of the climate prevents the culture of corn being anywhere successful. The aborigines, who are chiefly Samoides, maintain large herds of reindeer, and find their chief subsistence in the produce of fishing and of the chase. The town was formed in 1784 by the junction of the towns of Okladnikovo and Kouznetzova. It is situated about 162 miles north-east of Archangel, on the right bank of the Mezen, which here divides into two branches 18 miles above its entrance into the White Sea. The rivers are frozen from October to May. The only vegetables which are cultivated with any success are hemp and flax, of both of which great quantities are grown. The pastures are good, but neither horses, cattle, nor sheep are numerous. The forests are extensive. The wild animals are bears, wolves, foxes, ermines, and reindeer, with the birds common to such latitudes. Of the population, estimated in 1829 at 263,000, and in 1838 at 253,000, 5000 families of Samoides live between the Ural Mountains and the White Sea, and 2000 families of Lapps between the west coast of the White Sea and the Arctic Ocean. The inhabitants are largely employed in timber felling, and in the manufacture of charcoal, potash, and turpentine. The chief insular dependencies are Solovetzk, Waigatz, and Nova Zembla.

Lapland is divided by Wahlenburg into five zones, characterised by their vegetation. Professor Düben states that with regard to the extension of vegetation of different kinds, there may be distinguished eight different zones, in proceeding from the coast of the Gulf of Bothnia to the centre of the country, and from the lowlands to the mountain tops. The zone of the fir-tree, extending to 950 metres below the snow region, with a medium temperature of $+3°$ centigrade; the zone of the pine, extending to 831 metres below the snow region, with a medium temperature of +2·5°; the zone of the birch (594 metres), of the willow (416 metres), of the crowberry, *Empetrum nigrum* (236 metres), with a medium temperature of +1° centigrade; the Alpine zone, with spots of permanent snow; the zone of perpetual snow, extending from 920 to 1217 metres above the level of the sea, with a medium temperature of +0·4° centigrade; and lastly, the zone above this last. At 59 metres above the line, is the limit of vegetation.

The medium temperatures which have been stated indicate that cold must predominate throughout this vast region; but vegetation is on many spots very rich; the flowers there have an extraordinary brilliancy of colour, and vegetation developes itself with extreme rapidity. From the end of May the temperature may be very pleasant, rising to 20° centigrade towards the middle of the day. In the beginning of June occur *débâcles* in the lakes and rivers; by the 20th there are twenty-four hours of day, and the mean temperature of the month rises to 9·70°. July is very warm, with a mean temperature of 15·33°. By the 20th July barley is in ear; the hay-harvest occurs at the same time; and the plague of mosquitoes then attains its culmination. August is often very rainy, with a mean temperature of 15·36°. The harvest is generally terminated by the middle of August, some ten or twelve weeks after seed-time. Towards the middle of the month begin the long nights of autumn. In September the days are short, and this month is characterised by gales, accompanied by rain and snow. There is then made the gathering of wild berries, especially those of the cloud-berry (*Rubus chamæmorus*), which constitutes a very important article of food. The mean temperature is 5·40°. The other months belong to winter, with a mean temperature in October of 2·5°; in November of −1·98°; in December of −7·20°; in January of −17·50°; in February of −18·60°; in March of −11·40°; and in April of −3°.

A writer on Lapland in the *Edinburgh Encyclopædia* tells:—

'The temperature is remarkably similar throughout the whole extent of country between the Bothnian Gulf and the alpine ridge of mountains, about 69° of North Latitude. But in those parts which lie between the Lapland Alps and the Northern Ocean, the heat, excepting in some of the valleys, is almost entirely regulated by the latitude. In point of temperature, therefore, Lapland may be divided into two regions, the inland and the maritime. In the former the winter is very severe, and the summer very hot; in the latter the winter is comparatively mild, and the summer cold; the one being influenced by the temperature of the Frozen Ocean, and the other screened from its action by the alpine ridge forming a circle round it. The following table furnishes a comparative view of the mean temperature in both regions, by Fahrenheit's thermometer.

	At Enontekis, about 68¼ degrees, 1429 feet above the level of the sea.	At Mageroe, North Cape.
January,	0° 41'	22° 08'
February,	0 55	23 16
March,	11 41	24 71
April,	26 02	30 02
May,	36 56	34 07
June,	49 49	40 14
July,	59 63	46 42
August,	55 89	43 70
September,	41 78	37 62
October,	27 44	32 00
November,	12 20	25 75
December,	1 01	25 74
	26 85	32 13

'Though the mean temperature at Enontekis is nearly 6° lower than at the North Cape, yet is the former place better calculated for vegetation than the latter, and even brings to maturity certain kinds of corn, which is quite out of the question at the Cape. The reason is that the mean temperature during the summer months is much higher at Enontekis than at the Cape; and the power of vegetation is regulated more by the heat of summer than the cold of winter. In those countries, also, where the ground is long covered with snow, the temperature of the earth is considerably higher than that of the air, and this preserves it in a proper state for vegetation, in spite of the intense winter cold of the atmosphere. Thus, at Enontekis the ground is constantly covered with snow from the beginning of October to the beginning of May; while at the Cape, in consequence of the vicinity of the sea, it is more frequently exposed to thaws. Sometimes it happens in the Lapland Alps, that, when a colder summer than usual occurs, the snow lies during the whole year, and all kinds of vegetables are completely destroyed, except a few lichens, *Polytricha*, and *Peltidea crocea*. This is an event which occurs more frequently in Norwegian Lapland, where there are greater rains during summer, which reduce the temperature of the air, and prevent the dissolution of the snow, or even convert it into ice. The progress of the seasons may be readily perceived from the following table of observations made at Utsjocki, upon the river Tana, in 69° 53′ North Latitude.

Jan. 21. The sun's half-disc seen above the horizon.
May 5. First rain fell.
June 5. The ice disappeared upon the river Tana.
June 28. The lakes were free from ice.
July 15. Night frosts began.
Oct. 18. The rivers froze.
Oct. 25. The lakes froze.
Nov. 3. The ground covered with snow.
Nov. 20. The sun under the horizon.

'During the winter solstice, when the sun continues during seven weeks together under the horizon, instead of a clear daylight, there is only a twilight of a few hours. It is not so dark, however, but that a person might see to write, or do any ordinary business from ten o'clock in the forenoon to one o'clock in the afternoon; while the superior brightness of the moon and stars at this season, with the aid of the aurora borealis, and the reflection of the snow, supply in a great measure the absence of the sun. The cold, at this time of the year, is frequently so intense as to freeze brandy and spirits of wine. The lakes and rivers are covered with ice of extraordinary thickness; and the whole face of the country buried under snow to the depth of at least four or five feet. In the alpine regions the lakes have been known to be frozen to the depth of a fathom on the 9th of July; and the whole range of these mountains utterly impassable in winter, on account of the extreme cold; the total want of subsistence for the reindeer, and the violent gusts of wind which overturn both men and sledges. The drifting of the snow, when newly fallen, renders it impossible to go abroad till a partial thaw has taken place, when a hard crust is formed on the surface by frost, and enables the natives to travel on their sledges with the utmost celerity. During a thaw the atmosphere is surcharged with vapours; but when the north wind blows the air is clear and the sky beautifully serene. Thunderstorms are not uncommon even in the depth of winter. At the summer solstice the sun is as many weeks continually above the horizon as below it in winter; but his light during the night is paler, and less brilliant than during the day. The heat is then extremely oppressive, especially in the valleys; and the air is darkened by clouds of troublesome insects, which the natives have no possibility of avoiding, except by covering their heads with a cloth, or smearing their faces with tar, or involving themselves in the smoke of a fire. "The degree of heat," says Acerbi, "was twenty-nine (on the thermometer of Celsius) in the shade, and forty-five in the sun. The ground burned under our feet; and the few shrubs we met with in our way afforded us little or no shelter. We were almost suffocated with heat; and, to add to our sufferings, we were under the necessity of wearing a dress of thick woollen cloth as a security from the insects, and to cover our faces with a veil, which in a great measure prevented our drawing breath." In many parts of Lapland the days in summer are bright, serene, and warm, and the season, though short, remarkably healthy and delightful. At Altengaard, as observed by Baron Von Buch, in 79° North Lat. the thermometer generally stood at 70° or 72° in July; and the mean temperature of the month was nearly 63°.'

One consequence of the peculiarities of climatic condition is that most of the ports of the White Sea are frozen in winter, while Norwegian ports of a much higher latitude to the west of the White Sea remain open, and this notwithstanding the temperature on land there being lower than it is along the northern coast of Russia.

On this subject an interesting paper by Professor Daa, of the University of Christiana, was read at the International Congress of Students of Geographical Science, held in Paris in 1876. The following is a summary of this paper:—

It is generally known that the navigation of all the Russian ports on the White Sea is interrupted by ice during many months of the year, while the Norwegian coast remains open; and this remarkable difference has been attributed to the influence of the Gulf Stream, which moderates the climate of the country, but has no influence on that of the other. This opinion in regard to the contrast of climate in the two parts of the same sea is so rooted in the public opinion, that it is found in quite a number of publications, and yet, as a physical theory, it is destitute of any foundation. The frontier separating the two countries is altogether an artificial and arbitrary one. The Gulf Stream does not terminate at that point; it flows on to Nova Zembla, and it moderates in this way the climate of all these latitudes.

It is nevertheless the case that the Norwegian ports in the Arctic Ocean—Tromsoe, Hammerfest, Sandoe, and Vardoe —are never frozen during the winter, while the Russian towns of Kila, Kem, Onega, Archangel, Mezen, and Pustozersk, are shut off by the ice for many months every year.

In 1867, in a journey made along these coasts, Professor Daa traversed the interior from Kola to the White Sea and the Gulf of Bothnia, and had an opportunity of observing the causes of these apparently conflicting phenomena. These are sufficiently simple. The formation of ice on the surface of the ocean depends on the concurrence of many causes, amongst which the cold is one of but relative importance. Ice is formed more easily in inland basins, where the water is not so deep, is more mixed with fresh water, and is less exposed to great ocean waves. It is thus that the ice is formed in the Baltic, on the Zuyderzee, and sometimes even in the Adriatic. It is then natural that the interior parts of Norwegian fiords should also freeze. The port of Christiana, for example, is for some months closed by thick ice, which it is necessary to saw or break up by means of powerful steam vessels. On the contrary, on the margin of the ocean, the unceasing movement of the waves hinders congelation. Again, the interior waters offer the best protection for ships, and it is near to them that are found the most convenient positions for towns, and for communication with productive inland lands. The Russians, a people more especially agricultural, have built all their towns in interior localities, in order that communication may be had by rivers with their richest and most productive provinces. Now in inland seas the ice is formed for many miles on end, and navigation becomes impossible during winter. The Norwegians, who find their principal resources in navigation and fishing, have preferred building their towns on the shores of the ocean. The inconveniences thence resulting are many, and the ports are only of middling character. The Norwegian Assembly has been obliged to vote a sum of about two millions of francs to improve the port of Vardoe. But on the other hand, by nature or by art, it has been brought about that navigation can be carried on there continuously without interruption. It would, however, be erroneous to suppose that Norway constitutes an exception to the known laws of temperature, or that no natural obstacles to navigation exist there.

According to the opinion of Norwegian navigators, Russian Lapland possesses many excellent ports, for example, Tertik, the port of St. Catherine, and Kilden, but none of these are colonised or occupied by inhabitants living by the produce of the sea. The only town in Russian Lapland, Kola, has been built on the banks of a river, at about sixty kilometres from the ocean. It is not surprising then that this bay is covered with a sheet of unbreakable ice during many months of the year.

In fine, the difference of the winter in Russian Lapland and Norwegian Finmark is not the result of physical forces, but is attributable to differences in the habits of the two frontier nations.

CHAPTER II.

FLORA.

SECTION 1.—CHARACTERISTIC VEGETATION.

WE have found that the general appearance of the country is produced as much by its forests as by its general contour; but these, and the more lowly vegetation associated with them, may be with advantage brought under consideration apart, and the vegetation of the region throughout its several districts will be found to be regulated greatly by the climate, and more markedly so by the temperature, —only vegetables which can grow with little heat existing and dominating where the prevailing temperature is one adapted to their vegetation.

By Mackenzie Wallace it is stated :—'If it were possible to get a bird's-eye view of European Russia, the spectator would perceive that the country is composed of two halves widely differing from each other in character. The northern half is a land of forest and morass, plentifully supplied with water in the form of rivers, lakes, and marshes, and broken up by numerous patches of cultivation. The southern half is, as it were, the other side of the pattern—an immense expanse of rich arable land, broken up by occasional patches of sand or forest. The imaginary undulating line separating these two regions starts from the western frontier about the 50th parallel of latitude, and runs in a north-easterly direction till it enters the Ural Range at about 56 N. lat.' The northern half, however, he represents in a map illustrative of vegetation as divisible into two: the forest zone and the northern agricultural zone, the 60th parallel, or that of St. Petersburg, marking generally the line of division between them. It is the former alone with which we are here concerned; nor does the whole of it come under our cognisance. We have only to do with the forests in the Governments of Olonetz, Vologda, and Archangel, embracing the central and western portions of the zone.

With the forests of the Governments of Olonetz, Vologda, and Archangel, might be described the forests of the Governments of Viatka and Perm, but the exploitation of these, and more especially of the latter, is so affected by the demand for fuel used in connection with mining operations in the Ural Mountains, that with perhaps equal propriety they might be brought under consideration in connection with those of Eastern Russia. From the mention of this it will be seen that it is not in ignorance of this fact that they are not brought under consideration here.

It has been frequently remarked that if we note as we ascend a lofty mountain range the vegetation through which we pass, there are successive zones of these, varying in the kinds of plants by which they are characterised. If from the base of some lofty range of mountains in a tropical land, which, notwithstanding the high temperature in the plain, have their summits covered with perpetual snow, we ascend to this cooler region, we shall find vegetation of one kind giving place by degrees—tropical plants giving place to others, and these again to others, and such changes repeating themselves till at length we meet only with lichens and mosses and their allies. And like changes in the vegetation might be observed if we journeyed from the equator to either pole, representative of the successive zones on the mountain. Of this successive disappearance of different kinds of plants, as a mountain rises in altitude, Lapland supplies many illustrations. Baron von Buch writes:—

'It is extremely entertaining to climb great and rapidly

ascending heights in these climates. The vegetation with which we are familiar in the valleys gradually disappears under our feet. The Scotch fir soon leaves us; then the birches become shrivelled; now they wholly disappear; and between the bushes of mountain willows and dwarf birches, the innumerable clusters of berry-bearing herbs have room to spread—blae-berries on the dry heights, and mountain brambles on the marshy ground. We at last rise above them; the blae-berries no longer bear; they appear singly, with few leaves, and no longer in a bushy form. At last they disappear, and they are soon followed by the mountain willows. The dwarf birch alone braves the height and the cold; but at last it also yields before reaching the limit of perpetual snow; and there is a broad border before reaching this limit, on which, beside mosses, a few plants only subsist with great difficulty. Even the reindeer moss, which rises in the woods with the blaeberry in luxuriance of growth, is very unfrequent on such heights. On the top of the mountains, which is almost a table-land, there is no ice, it is true, nor glaciers; but the snow never leaves these heights; and a few single points and spots above the level are alone clear of snow for a few weeks. It is a melancholy prospect; nothing in life is any longer to be seen, except perhaps occasionally an eagle in his flight over the mountains from one fiord to another.'

On Akka Solki, one of these mountains on the western coast, which is about 3392 English feet in height, the following limits of the different productions were accurately marked:—

	Eng. Feet.
Limit of snow in latitude 70°,	3514
Betula nana, or dwarf birch,	2742
Salix myrsinitis, or whortle-leaved willow,	2150
Salix lanata, or downy willow, rises above the *Betula nana*, and approaches the limit of perpetual snow.	
Vaccinium myrtillus, or blae-berry.	2031
Betula alba, or birch tree,	1579

We should find following each other in the same order, but in broader zones, in the tropical, sub-tropical, temperate, and arctic or antarctic regions of the globe, many of these zones being susceptible of well-defined sub-divisions.

According to a report published anonymously, but attributed to Admiral Count Mordvinoff, Director of the Agricultural Society of Russia some fifty years ago, who laboured assiduously to develope the agriculture of his country, Northern Russia comprises four well defined regions.

The first is the region of ice. The icy region may be considered as including Nova Zembla, or more correctly, *Novaya Zembia*, or New Land, part of the Kolskaya district, and the extreme northern point of land which projects into the Frozen Ocean. This region is distinguished by a night of three months' duration, and by its total destitution of vegetable productions, which circumstances render it unfit for the permanent habitation of man and domestic animals. The seal, the walrus, and fish of various descriptions, which abound toward the pole, supply the only means of sustenance for man, the Polar bear, and its inseparable companion, the fox, except on Novaya Zembia, where multitudes of a peculiar kind of mice breed, and lay up heaps of roots for their winter store, and these mice serve in their turn as food for the bears and foxes.

The second is the region of moss. The mossy region, where the ever-frozen ground is covered with a kind of greyish moss, and towards the boundaries of the following region with a kind of brushwood and fir. This tract is endowed by nature with an animal which alone makes it habitable for man—the reindeer. Its vast deserts stretching from Archangel along the shores of the White Sea to the Eastern Ocean, are peopled by thinly scattered nomadic tribes of Laplanders, Samoyeds, Ostiaks, and other aborigines, whose numbers are gradually decreasing as they come in contact with civilised nations. . . . . In this region, adjacent to the Frozen Ocean, at the mouth of great rivers, and near certain islands, are found astonishing remains of antediluvian animals, particularly of the mammoth; and here were discovered the bones of one of these monsters still covered with flesh and skin.

To this there succeeds the region of forest and pasturage. By degrees the dwarf trees and brushwood of the mossy region increase in size, until we come to those immense forests, where the hand of man has scarcely as yet disturbed the majestic operations of nature. Along the banks of the rivers, and in other spots unencumbered with wood, the grass shoots up with astonishing rapidity; but the lingering frosts of spring, and the early appearance of those of autumn prevent the cultivation of corn. For this reason the inhabitants of the northern part of this district are principally occupied with the chase, especially that of the squirrel, an animal which seems to be indigenous there, and which forms the principal inducement for man to take up his abode in this inhospitable clime. The abundance of grass in the southern parts affords the means of keeping cattle, while in some sheltered spots appear a few corn-fields, as it were the out-posts of agriculture. The northern and eastern parts of this region are inhabited by nomadic tribes, then follow the Finns or Finlanders, a settled people, chiefly dependent on pasturage for support. It would be difficult to mark with precision the southern boundaries of this region, as it falls gradually into the next.

This is designated by him 'the region of barley and the beginning of agriculture.' The inhabitants of the region extending beyond as well as within that portion of Russia in Europe, which it includes, are Russians, Finns, Zirians, and others having settled habitations; but, from the insignificance of their agriculture, they have recourse to grazing, fishing, and the chase, the floating of timber, &c.; and in some parts of the Governments of Archangel and Vologda are to be found a very superior breed of horned cattle. The southern limits of this region may be said to extend nearly to the town of Yarensk, in the Government of Vologda, and the parts of a corresponding degree of latitude, viz., 63°. Nature, as the author himself had an opportunity of observing, here assumes an imposing aspect—immense forests, vast rivers, beautiful meadows flourishing in all the unexhausted luxuriance of primitive vegetation, make an impression on the traveller that can only be adequately conceived by those who have wandered through the unexplored forests, and beheld the majestic streams of the New World.

Thus do we pass from the Arctic Circle into the temperate zone, finding each stage marked by a change in the character of the vegetation. And the vegetation characteristic of the several regions which have been so defined may with interest be studied in its details.

It will be found that even the icy region of the Arctic Circle is not without its vegetation; marine *algæ* of more or less complex structure may be considered the primary food of all organisms in that region belonging to the animal kingdom. Here, as in the study of fossils, the existence of animals may be accepted as indicative of the existence at the same time and place of vegetable organisms; and here the abundant *fauna* speaks of a most abundant flora.

With regard to the wide dispersion of fish, both in the northern portion of the temperate zone and within the Arctic Circle, it is stated by the author of *The Arctic World: its Plants, Animals, and Natural Phenomena**—

'The wealth of the Arctic and sub-Arctic seas is apparently inexhaustible. In many parts cod are plentiful, and supply the Greenlanders with a valuable article of food. The capelin (*Mallotus villosus*), which in May and June frequents the Greenland waters, is eaten both fresh and dried; in the latter case forming a useful winter provision. The halibut is found of a large size; and ocean also contributes the Norway haddock, the salmon-trout,

---

* London: T. Nelson & Sons.

the lump-fish, and the bull-head. Nor are the crustacea unrepresented; long-tailed crabs being abundant, while the common mussel may be gathered almost everywhere at ebb-tide. The seas, however, grow poorer as we advance towards the Pole, and many important species of fish do not penetrate further north than the Arctic Circle.

'Yet even where these are wanting, the ocean-waters teem with life; and a recent writer is fully justified in remarking that the vast multitudes of animated beings which people them form a remarkable contrast to the nakedness of their bleak and desolate shores. The colder surface-waters are, as he says, almost perpetually exposed to a cold atmosphere, and being frequently covered, even in summer, with floating ice, they are not favourable to the development of organic life; but this adverse influence is modified by the higher temperature which constantly prevails at a greater depth. Contrary to the rule in the Equatorial seas, we find in the Polar ocean an increase of temperature from the surface downwards, in consequence of the warmer under-currents, flowing from the south northwards, and passing beneath the cold waters of the superficial Arctic current.

'Hence the awful rigour of the Arctic winter, which strikes the earth with a death-blight, is not perceptible in the ocean-depths, where myriads of organisms find a secure retreat from the frost, and whence they emerge during the long summer's day, either to haunt the shores or ascend the broad rivers of the Polar world. Between the parallels of 74° and 80°, Dr Scoresby observed that the colour of the Greenland sea varies from the purest ultramarine to olive-green, and from crystalline transparency to striking opacity; and these appearances are not transitory, but permanent.* The aspect of this green semi-opaque water, which varies in its locality with the currents—often forming isolated stripes, and sometimes spreading over two

---

* Scoresby calculated that it would require 80,000 persons, labouring continuously from the creation of man to the present day to count the number of organisms contained in two miles of the green water.

or three degrees of latitude—is mainly due to small medusæ and nudibranchiate molluscs Many thousands of square miles must literally run riot with life, since the coloured waters we speak of are calculated to form one-fourth of the sea between the 74th and 80th parallels.'

All of these animated beings tell of vegetation, for this supplies the primary food by which they are sustained. It is so on land: there we have herbivori and carnivori animals feeding on the vegetable productions of the earth, and beasts and birds, and insects innumerable which prey upon these, but they, too, are sustained by the grass and the herb of the field, for by these have their prey been nourished, and like to life on earth is life on the sea.

'On the Greenland coast,' says in continuation the writer I have quoted, 'where the transparency of the waters is so great that the bottom and every object upon it are clearly discernible, even at a depth of eighty fathoms, the ocean-bed is covered with gigantic tangles, so as to remind the spectator of the ocean-gardens of the Tropical Zone. Alcyonians, sertularians, acidians, nullipores, mussels, and a variety of other sessile animals incrust every stone, or congregate in every fissure and hollow of the rocky ground. A dead seal or fish flung into the sea is soon converted into a skeleton, it is said, by the myriads of small crustaceans which infest these northern waters, and, like the ants in the equatorial forests, perform the part of scavengers of the deep.'*

---

* He adds:—'It is evident from the observations of Professor Forbes, that *depth* has a very considerable influence in the distribution of marine life. From the surface to the depth of 1380 feet eight distinct zones or regions have been mapped out in the sea, each of which has its own vegetation and inhabitants; and the number of these regions must now be increased, after the astonishing results of the deep-sea soundings of Dr Carpenter and Professor Wyville Thomson. The changes in the different zones are not abrupt; some of the creatures of an under region always appear before those of the region above it vanish; and though there are a few species the same in some of the eight zones, only two are common to all. It is to be observed that those near the surface have forms and colours analogous to the inhabitants of southern latitudes, while those at a greater depth are analogous to the animals of northern waters. Hence, in the sea, *depth* corresponds with latitude, as *height* does on land. Mrs Somerville adds, in language of much terseness, that the extent of the geographical distribution of any species is proportioned to the depth at which it lives. Consequently, those which live near the surface are less widely dispersed than those inhabiting deep water.

'The larger and more active inhabitants of the seas obey the same laws with the rest of creation, though their provinces, or regions, are in some instances very extensive.'

But it is the land with which we are more concerned, and even within the limits of perpetual snow there is found upon the surface of the frozen covering of the earth and sea a minute vegetable organism, which was early designated, in ignorance of its structure, 'Red Snow.'

'This so-called "red snow" says the author of *The Arctic World*, was found by Sir John Ross, in his first Arctic expedition in 1808, on a range of cliffs rising about 800 feet above the sea-level, and extending eight miles in length (lat. 75° N.) It was also discovered by Sir W. E. Parry in his overland expedition in 1827. The snow was tinged to the depth of several inches. Moreover, if the surface of the snow-plain, though previously of its usual spotless purity, was crushed by the pressure of the sledges and of the footsteps of the party, blood-like stains instantly arose; the impressions being sometimes of an orange hue and sometimes more like a pale salmon tint.

'It has been ascertained that this singular variation of colour is due to an immense aggregation of minute plants of the species called *Protococcus nivalis*; the generic name alluding to the extreme primitiveness of its organisation, and the specific to the peculiar nature of its habitat. If we place a small quantity of red snow on a piece of white paper, and allow it to melt and evaporate, there will be left a residuum of granules sufficient to communicate a faint crimson tint to the paper. Examine these granules under a microscope, and they will prove to be spherical purple cells of almost inappreciable size, not more than the three-thousandth to one-thousandth part of an inch in diameter. Look more closely, and you will see that each cell has an opening, surrounded by indented or serrated

---

Above the 44th parallel the Atlantic species frequently correspond with those of the Pacific. The salmon of America is identical with that of the British Isles, and the coasts of Sweden and Norway; the same is true of the *Gadidae*, or cod. The *Cottas*, or bull-head tribe, are also the same on both sides of the Atlantic; increasing in numbers and specific differences on approaching the Arctic seas. The same law holds good in the North Pacific, but the *generic* forms differ from those in the Atlantic. From the propinquity of the coasts of America and Asia at Behring Strait, the fish on both sides are nearly alike, down to Admiralty Inlet on the one side, and the Sea of Okhotsk on the other.'

lines, the smallest diameter of which measures only the five-thousandth part of an inch. When perfect, the plant, as Dr Macmillan observes, bears a resemblance to a redcurrant berry; as it decays, the red colouring matter fades into deep orange, which is finally resolved into a brownish hue. The thickness of the wall of the cell is estimated at the twenty-thousandth part of an inch, and three hundred to four hundred of these cells might be grouped together in a smaller space than a shilling would cover. Yet each cell is a distinct individual plant; perfectly independent of others with which it may be massed: fully capable of performing for and by itself all the functions of growth and reproduction; possessing "a containing membrane which absorbs liquids and gases from the surrounding matrix or elements, a contained fluid of peculiar character formed out of these materials, and a number of excessively minute granules equivalent to spores, or, as some would say, to cellular buds, which are to become the germs of new plants." Dr Macmillan adds: "That one and the same primitive cell should thus minister equally to absorption, nutrition, and reproduction, is an extraordinary illustration of the fact that the smallest and simplest organised object is in itself, and, for the part it was created to perform in the operations of nature, as admirably adapted as the largest and most complicated."

'The first vegetable forms to make their appearance at the limits of the snow-line, whether in high latitudes or on mountain-summits, are lichens; which flourish on rocks, or stones, or trees, or wherever they can obtain sufficient moisture to support existence. Upwards of two thousand four hundred species are known. The same kinds prevail throughout the Arctic regions, and the species common to both the Eastern and Western hemispheres are very numerous. They lend the beauty of colour to many an Arctic scene which would otherwise be inexpressibly dreary; the most rugged rock acquiring a certain air of picturesqueness through their luxuriant display. Their forms are wonderfully varied; so that they present to the student of Nature an almost inexhaustible field of inquiry. In their most rudimentary aspects they seem to consist of nothing more than a collection of powdery granules, so minute that the figure of each is scarcely distinguishable, and so dry and so deficient in organisation that we cannot but wonder how they live and maintain life. Now they are seen like ink-spots on the trunks of fallen trees; now they are freely sprinkled in white dust over rocks and withered tufts of moss; others appear in gray filmy patches; others again like knots or rosettes of various tints; and some are pulpy and gelatinous, like aërial sea-weeds which the receding tide leaves bare and naked on inland rocks. A greater complexity of structure, however, is visible in the higher order of lichens—and we find them either tufted and shrubby, like miniature trees; or in clustering cups, which, Hebe-like, present their " dewy offerings to the sun."

'In the Polar world, and its regions of eternal winter, where snow and ice, and dark drear waters, huge glacier and colossal berg, combine to form an awful and impressive picture, the traveller is thankful for the abundance of these humble and primitive forms, which communicate the freshness and variety of life to the otherwise painful and death-like uniformity of the frost-bound Nature.

'A lichen which is discovered in almost every zone of altitude and latitude, which ranges from the wild shores of Melville Island in the Arctic to those of Deception Island in the Antarctic circle,—which blooms on the crests of the Himalayas, on the lofty peak of Chimborazo, and was found by Agassiz near the top of Mount Blanc,—is the *Lecidea geographica*, a beautiful bright-green lichen, whose clusters assume almost a kaleidoscopic appearance.

'A lichen of great importance in the Arctic world is the well-known *Cladonia rangiferina,* or reindeer moss, which forms the staple food of that animal during the long Arctic winter. In the vast tundras, or steppes, of Lapland, it flourishes in the greatest profusion, completely covering the ground with its snowy tufts, which look like the silvery sprays of some magic plant. According to Linnæus, it thrives more luxuriantly than any other plant in the pine-forests of Lapland, the surface of the soil being carpeted with it for many miles in extent; and if the forests are accidentally burned to the ground, it quickly reappears, and grows with all its original vigour.

'When the ground is crusted with a hard and frozen snow, which prevents it from obtaining its usual food, the reindeer turns to another lichen, called rock-hair (*Alectoria jubata*), that grows in long bearded tufts on almost every tree. In winters of extreme rigour the Laplanders cut down whole forests of the largest trees, that their herds may browse freely on the tufts which clothe the higher branches. Hence it has been justly said that "the vast dreary pine-forests of Lapland possess a character which is peculiarly their own, and are perhaps more singular in the eyes of the traveller than any other feature in the landscapes of that remote and desolate region. This character they owe to the immense number of lichens with which they abound. The ground, instead of grass, is carpeted with dense tufts of the reindeer moss, white as a shower of new fallen snow; while the trunks and branches of the trees are swollen far beyond their usual dimensions with huge, dusky, funereal branches of the rock-hair, hanging down in masses, exhaling a damp earthy smell, like an old cellar, or stretching from tree to tree in long festoons, waving with every breath of wind, and creating a perpetual melancholy sound."

'In the regions furthest north are found various species of lichens belonging to the genera *Gyrophora* and *Umbilicaria*, and known in the records of Arctic travel as rock tripe, or *tripe de roche*; a name given to them in consequence of their blistered thallus, which bears a faint resemblance to the animal substance so called. They afford a coarse kind of food, and proved of the greatest service to the expeditions under Sir John Franklin; though their nutritious properties are not considerable, and, such as they are, are unfortunately impaired by the presence of a bitter principle which is apt to induce diarrhœa. In Franklin and Richardson's terrible overland journey from the Coppermine River to Fort Enterprise it was almost the sole support, at one time, of the heroic little company. Dr Richardson says they gathered four species of *Gyrophora*,* and used them all as articles of food; "but not having the means of extracting the bitter principle from them, they proved nauseous to all, and noxious to several of the party, producing severe bowel complaints." Franklin on one occasion remarks: "This was the sixth day since we had enjoyed a good meal; the *tripe de roche*, even when we got enough, only serving to allay the pangs of hunger for a short time." Again, we read: "The want of *tripe de roche* caused us to go supperless to bed."

'Dr Hayes, in the course of his "Arctic Boat Journey," was compelled to have recourse to the same unsatisfactory fare. The rock-lichen, or stone-moss, as he calls it, he describes as about an inch in diameter at its maximum growth, and of the thickness of a wafer. It is black externally, but when broken the interior appears white. When boiled it makes a glutinous fluid, which is slightly nutritious.

'"Although in some places it grows very abundantly," writes Dr Hayes, "yet in one locality it, like the game, was scarce. Most of the rocks had none upon them; and there were few from which we could collect as much as a quart. The difficulty of gathering it was much augmented by its crispness, and the firmness of its attachment.

'"For this plant, poor though it was, we were compelled to dig. The rocks in every case were to be cleared from snow, and often our pains went unrewarded. The first time this food was tried it seemed to answer well,—it at least filled the stomach, and thus kept off the horrid sen-

---

* So called from its circular form, and because the surface of the leaf is marked with curved lines.

sation of hunger until we got to sleep; but it was found to produce afterwards a painful diarrhœa. Besides this unpleasant effect, fragments of gravel, which were mixed with the mosses, tried our teeth. We picked the plants from the rock with our knives, or a piece of hoop-iron; and we could not avoid breaking of some particles of the stone."

'These lichens are black and leather-like, studded with small black points like "coiled wire buttons," and attached either by an umbilical root or by short and tenacious fibres to the rocks. Some of them may be compared to a piece of shagreen, while others resemble a fragment of burned skin. They are met with in cold bleak localities, on Alpine heights of granite or micaceous schist, in almost all parts of the world,—on the Scottish mountains, on the Andes, on the Himalayas; but it is in the Polar world that they most abound, spreading over the surface of every rock a sombre Plutonian vegetation, that seems to have been scathed by fire and flame, until all its beauty and richness were shrivelled up.'

In Sweden, while several of the lichens produced in these regions are employed as dyes, others are used as medicine, and some as poisons for noxious and dangerous animals which are found annoying.

'The only lichen which has retained its place in modern pharmacy is the well-known "Iceland moss." It is still employed as a tonic and febrifuge in ague; but more largely, when added to soups and chocolate, as an article of diet for the feeble and consumptive. In Iceland the *Cetraria Islandica* is highly valued by the inhabitants. What barley, rye, and oats are to the Indo-Caucasian races of Asia and Western Europe; the olive, the fig, and the grape to the inhabitants of the Mediterranean basin; rice to the Hindu; the tea-plant to the native of the Flowery Land; and the date-palm to the Arab,—is Iceland moss to the Icelander, the Lapp, and the Eskimo.

It is found on some of the loftiest peaks of the Scottish

Highlands; but in Iceland it overspreads the whole country, flourishing more abundantly, and attaining to a larger growth on the volcanic soil of the western coast than elsewhere. It is collected triennially, for it requires three years to reach maturity, after the spots where it thrives have been cleared. We are told that the meal obtained from it, when mixed with wheat-flour, produces a greater quantity, though perhaps a less nutritious quality, of bread than can be manufactured from wheat-flour alone. The great objection to it is its bitterness, arising from its peculiar astringent principle, cetraria. However, the Lapps and Icelanders remove this disagreeable pungency by a simple process. They chop the lichen to pieces, and macerate it for several days in water mixed with salt of tartar or quicklime, which it absorbs very readily; next they dry it, and pulverize it; then, mixed with the flour of the common knot-grass, it is made into a cake, or boiled, and eaten with reindeer's milk.

'Mosses are abundant in the Arctic regions, increasing in number and beauty as we approach the Pole, and covering the desert land with a thin veil of verdure, which refreshes the eye and gladdens the heart of the traveller. On the hills of Lapland and Greenland they are extensively distributed; and the landscape owes most of its interest to the charming contrasts they afford. Of all the genera, perhaps the bog-mosses, *Sphagna*, are the most luxuriant; but at the same time they are the least attractive, and the plains which they cover are even drearier than the naked rock. In Melville Island these mosses form upwards of a fourth part of the whole flora. Much finer to the sight is the common hair-moss (*Polytrichum commune*), which extends over the levels of Lapland, and is used by the Lapps, when they are bound on long journeys, for a temporary couch. We may mention also the fork-moss (*Dicranum*), which the Eskimos twist into wicks for their rude lamps.

'Of other cryptogams fungi extend almost to the very limits of Arctic vegetation. The Greenlanders and Lapps make use of them for tinder, or as styptics for stopping the flow of blood, and allaying pain. In Siberia they abound. Frequently, in the high latitudes, they take the form of "snow mould," and are found growing on the barren and ungenial snow. These species are warmed into life only when the sun has grown sufficient to melt the superficial snow-crust, without producing a general thaw, and then they spread far and wide in glittering wool-like patches, dotted with specks of red or green. When the snow melts they overspread the grass beneath like a film of cob-web, and in a day or two disappear.

'During Captain Penny's voyage in search of Sir John Franklin he picked up two pieces of floating drift-wood, far beyond the usual limit of Eskimo occupation, which, from their peculiar appearance excited a lively curiosity. The one was found in Robert Bay, off Hamilton Island, lat. 76° 2' north, and long. 76° west—that is, in the route which Franklin's ships, it is supposed, had followed,—and was plainly a fragment of wrought elm plank, which had been part of a ship's timbers. It exhibited three kinds of surface—one that had been planed and pitched, one roughly sawn, and the third split with an axe. The second piece of drift-wood was picked up on the north side of Cornwallis Island, in lat. 75° 36' north, and long. 96° west. It was a branch of white spruce, much bleached in some places, and in others charred and blackened as if it had been used for fuel.

'On both fragments traces of microscopic vegetation were discovered; and as it was thought they might, if carefully examined, afford some clue to the fate of Franklin's expedition, they were submitted to Mr Berkeley, a well-known naturalist. In the report which he addressed to the Admiralty he stated that the vegetation in both cases resembled the dark olive mottled patches with which wooden structures in this country, if exposed to atmospheric influences, are speedily covered. The bleached cells and fibres of the fragment of elm were filled up with slender fungoid forms, *mycelia;* while on its different surfaces appeared several dark-coloured specks, belonging to the genus *Phoma.* As it was not probable that plants so minute could have retained, through the terrible severity of an Arctic winter, their delicate naked spores in the perfect condition in which they were found, Mr Berkeley concluded that they must have been developed through that same summer; while from three to four years, in those high latitudes, and amid the rigour of stormy ice-covered seas, would suffice to produce the bleached appearance of the wood. Hence he inferred that the plank had not been long exposed.

'On the other fragment of drift-wood he discovered some deeply-embedded minute black fungoid forms, called *Sporidesmium lepraria.* Unlike the *phomas,* which are very ephemeral, these plants possess the longevity of the lichens, and the same patches last for years unchanged on the same pieces of wood, while their traces are discernible for a still longer period. From their condition Mr Berkeley concluded that the fungi on the drifted wood had not been recently developed, but that, on the contrary, they were the remains of the species which existed on the drift-wood when used for fuel by the unfortunate crews of Franklin's ships, the *Erebus* and the *Terror.*

'There can be no doubt whatever, as Dr Macmillan remarks, considering the circumstances in which they were discovered, and the remarkable appearances they presented —there can be no reasonable doubt that both fragments of drift-wood belonged to, or were connected with, the lost ships; and the curious information regarding the course they pursued at a certain time, furnished by witnesses so extraordinary and unlikely as a few tiny dark specks of cryptogamic vegetation on floating drift-wood, was confirmed, in a wonderful manner by the after-discovery of the first authentic account ever obtained of the sad and pathetic history of Franklin's expedition.*

---

* In Siberia grows the fly-agaric (*Agaricus muscarius*), from which the inhabitants obtain an intoxicating liquor of peculiarly dangerous character. It has a tall white

By Arctic voyagers mention is occasionally made of the scurvy grass, a species of *Cochlearia*, of the medicinal properties of which some of the undaunted explorers of the Polar regions were fain to avail themselves.

The shores of the Polar Sea, with a severe temperature modified by proximity to the ocean, are not destitute of terrestrial vegetation. In Spitzbergen, with a latitude of 77° to 81°, besides delicate mosses which clothe the moist low lands, and hardy lichens which encrust the rocks up to the remotest limits of vegetation, and which are very numerous, there have been found about ninety-three species of phanerogamous, or flowering plants, amongst which are the *Arenaria biflora*, the *Cerastium alpinum*, and the *Ranunculus glacialis*, found on the Alps at altitudes varying from 9,000 to 10,000 feet above the level of the sea. The only esculent plant is the *Cochlearia fenestrata*, which here is devoid of the bitterness of which Arctic explorers complain, and may be eaten as a salad; and several grasses supply, along with the Iceland moss, food for the reindeer.

And thus it is also with the dreary wastes of the tundras. 'Though not rich in bud and bloom, even these dreary wastes are not absolutely without floral decoration. Selinum and cerastium, as well as the poppy and sorrel, andromeda, and several species of heath, are mentioned by Dr Kane as blooming in the neighbourhood of Smith Strait. On the south coast of the Polar Sea Dr Richardson found a considerable variety of vegetation.'

---

stem, surmounted by a dome of rich orange scarlet, studded with white scaly tuberules, and in some parts of Kamtschtka and the northern districts of Siberia is so abundant that the ground sparkles and shines as if covered with a scarlet carpet. The natives collect it during the hot summer months, and dry it. Steeped in the juice of the whortleberry, it forms a powerful intoxicating wine; or rolled up like a bolus, and swallowed without chewing, it produces much the same effect as opium. On some, however, it acts as an excitant, and induces active muscular exertion. A talkative person, under its influence, cannot keep silence or secrets; one fond of music, sings incessantly; and if a person who has partaken of it wishes to step over a straw or small stick, he takes a stride or jump sufficient to clear the trunk of a tree!

The Koriaks and Kamtschatkans personify this fungus, under the name of *Moche Moro*, as one of their *penates*, or household gods; and if they are impelled by its effects to commit any dreadful crime, they pretend they act only in obedience to commands which may not be disputed. To qualify themselves for murder or suicide, they drink additional doses of "this intoxicating product of decay and corruption."

'We noticed,' he says, 'about one hundred and seventy phanerogamous or flowering plants; being one-fifth of the number of species which exist fifteen degrees of latitude further to the southward. He adds: The grasses, bents, and rushes, constitute only one-fifth of the species on the coast, but the two former tribes actually cover more ground than all the rest of the vegetation. The *cruciferae*, or cross-like tribe, afford one-seventh of the species, and the compound flowers are nearly as numerous. The shrubby plants that reach the sea-coast are the common juniper, two species of willow, the dwarf birch, the common alder, the hippophaë, the gooseberry, the red bear berry (*arbutus uva ursi*), the Labrador tea-plant, the Lapland rose, the bog-whortleberry, and the crowberry. The kidney-leaved oxyria grows in great abundance there, and occasionally furnished us with an agreeable addition to our meals, as it resembles the garden-sorrel in flavour, but is more juicy and tender. It is eaten by the natives, and must, as well as many of the cress-like plants, prove an excellent corrective of the gross, oily, rancid, and frequently putrid meat on which they subsist. The small balls of the Alpine bistort, and the long, succulent, and sweet roots of many of the astragaleæ, which grow on the sandy shores, are eatable; but it does not seem that the Eskimos are acquainted with their use. A few clumps of white spruce-fir, with some straggling black spruces and canoe-birches, grow at the distance of twenty or thirty miles from the sea, in sheltered situations on the banks of rivers.

'It has been pointed out that the principal characteristic of the vegetation of the Arctic regions is the predominance of perennial and cryptogamous plants; but further southward, where night begins to alternate with day, or in what may be called the sub-arctic zone, a difference of species appears which greatly enhances the beauty of the landscape. A richly and vividly-coloured flora adorns these latitudes in Europe as well as in Asia during their brief but ardent summer, with its intense radiance and intense warmth,—consisting of potentillas, gentians, starry chickweeds, spreading saxifrages and sedums, spiræas, drabas, artemisias, and the like. The power of the sun is so great, and the consequent rapidity of growth so extraordinary, that these plants spring up, and blossom, and germinate, and perish, in six weeks. In a lower latitude many ligneous plants are found,—as berry-bearing shrubs, the glaucous kalmia, the trailing azalea, the full-blossomed rhododendron. The Siberian flora differs from the European in the same latitudes by the inclusion of the North American genera, phlox, mitella, and claytonia, and by the luxuriance of its asters, spiræas, milk-vetches, and the saline plants goosefoot and saltwort.

'In Nova Zembla, and other northern regions, the vegetation is so stunted that it barely covers the ground, but a much greater variety of minute plants of considerable beauty are aggregated there in a limited space than in the Alpine climes of Europe where the same genera occur. This is due to the feebleness of the vegetation; for in the Swiss Alps the same plant frequently usurps a large area, and drives out every other,—as the dark-blue gentian, the violet-tinted pansy, and the yellow and pink stone-crops. But in the far north, where vitality is weak and the seeds do not ripen, thirty different species, it has been observed, may be seen "crowded together in a brilliant mass," no one being powerful enough to overcome its companions. In these frozen climates plants may be said to live between the air and the earth, for they scarcely raise their heads above the soil, and their roots, unable to penetrate it, creep along the surface. All the woody plants—as the betula nava, the reticulated willow, andromeda tetragona, with a few bacciferous shrubs—trail upon the ground, and never rise more than an inch or two above it. The *Salix lanata*, the giant of the Arctic forests, is about five inches in height; while its stem, ten or twelve feet long, lies hidden among the moss, and owes shelter, almost life, to its humble neighbour.

'In the wooded zone the thermometer does not rise

above zero until the month of May. Then, under the influence of a more genial temperature, the breath of life passes into the slumbering, inert vegetation. Then the reddish shoots of the willows, the poplars, and the birches, hang out their long cottony catkins; a pleasant greenness spreads over copse and thicket; the dandelion, the burdock, and the saxifrages lift their heads in the shelter of the rocks; the sweet-brier fills the air with fragrance, and the gooseberry and the strawberry are put forth by a kindly nature; while the valleys bloom and the hill-sides are glad with the beauty of the thuja, the larch, and the pine.

'On the southern margin of the wooded region, as in Sweden, Russia, and Siberia, extend immense forests, chiefly of coniferous trees. As we move towards the north, these forests dwindle into scattered woods and isolated coppices, composed chiefly of stunted poplars and dwarf birches and willows. The sub-alpine myrtle and a small creeping honeysuckle with rounded leaves are met with in favourable situations. Continuing our northerly progress, we wholly leave behind the arborescent species; but the rocks and cliffs are bright with plants belonging to the families of the ranunculaceæ, saxifragaceæ, cruciferæ, and gramineæ. To the dwarf firs and pigmy willows succeed a few scattered shrubs—such as the gooseberry, the strawberry, the raspberry, pseudo-mulberry (*Rubus chamæmorus*)—indigenous to this region, and the Lapland oleander (*Rhododendron laponicum.*)

'Still advancing northward, we find, at the extreme limits of the mainland, some drabas (*Cruciferae*), potentillas (*Rosaceae*), burweeds and rushes (*Cyperaceae*), and lastly a great abundance of mosses and lichens. The commonest mosses are the *Splechnum*, which resembles small umbels; and, in moist places, the *Sphagnum*, or bog-moss, whose successive accumulations, from a remote epoch, have formed with the detritus of the *Cyperaceae*, extensive areas of peat, which at a future day will perhaps be utilised for fuel.'*

---

* *The Arctic World:* Its Plants, Animals, and Natural Phenomena. London: T. Nelson & Sons. 1876.

Section II.—Forests.

The forests in the central portion of Archangel may be considered as bounded on the north by the Arctic circle, but on the north-western extremity of the Government they extend a considerable way beyond that circle, and on the north-eastern part of the Government they stretch, by a tapering projection, still further within the circle. Beyond this curved boundary we meet only with dwarf trees and brushwood, and fewer and fewer of these till only moss is seen, and this at length covered extensively with snow, till finally we reach the region of perpetual snow enclosed within the embrace of a frozen ocean.

In the Government of Archangel there is an area of 30,312,002 deciatins of forests, of which 38,549,270 *dec.* belong to the Crown, equivalent to 46 *dec.* of forests, or 43.2 *dec.* of Crown forests to every square verst. The annual fellings in the Crown forests yield 1.2 cubic feet, and the revenue is 0.2 kopecs per deciatin.

In the Government of Olonetz the area of forests is 11,461,00 deciatins, of which 10,497,250 *dec.* belong to the Crown, equivalent to 99.6 *dec.* of forest, or 91.2 *dec.* of Crown forests per square verst. The annual fellings in the Crown forests yield 1.2 cubic feet, and the revenue 0.2 kopecs per deciatin.

In the Government of Vologda there are 33,470,000 deciatins of forests, of which 39,346,037 belong to the Crown, equivalent to 95 *dec.* of forest, or 86.2 *dec.* of Crown forests per square verst. The annual fellings in the Crown forests yield 2.7 cubic feet, and the revenue is 1.1 kopec per deciatin.

A description of one of the forests, superadded to the descriptions which have been given of the forest lands of these Governments, may suffice to convey a general idea what may be seen in these.

Of the forest estate of Vuig Mr Judrae gives the following account:—'The Vuig forest estate covers an area of 530,000 deciatins. It lies to N.N.E. of the district of Povonetz, and in its northern part it abuts on the Government of Archangel. Within its area there are about 2100 inhabitants of both sexes, so that there are more than 250 deciatins for each person. The soil of this estate is sand; in some places it is damp and covered with a deep layer of moss; and there are a great many bogs, marshes, and small lakes in low-lying places; but withal there are alongside of these dry meadows and undulating ridges, and even hills of rock, consisting of granite and other primitive formations, partly covered with drift-sand and partly bare; stones cover the ground everywhere. The climate is very severe; there is scarcely any spring; sometimes in the end of May the lakes and rivers are free of ice; and after that the heat sets in with scarcely an intervening period.

'The absence of darkness during the summer, and, in consequence of this, the greater and continuous action of the solar rays, allows of continuous vegetation; and growth proceeds very rapidly. But even at mid-summer, with the constant north winds, there are frequent frosts. The grain ripens in August, and by the end of that month autumn begins. The first snow falls generally in October, but sometimes it does not till November; and not till this month, and sometimes not till the month of December, do the lakes freeze. The lateness of the freezing of the lakes is attributed to the autumn being warm in places adjacent to the ocean. The prevailing winds, partly in autumn and in winter, are northern: the midnight or mid-winter wind blowing from the N.N.E. These winds are very constant and continuous. In consequence of the abundance of waters and woods, the difference between the temperature of the day and night is very considerable in summer. Frequently after a hot day, there is so much frost during the night that in the marshes the water freezes.

'All of these peculiarities, characteristic as they are of the soil and of the climate, could not fail to have an influence on vegetable and animal life there, and give an impress to the occupations, manners, and mode of life of the population.

'With animal life in this country I am but little acquainted, having given little attention to zoology; but researches devoted to the *fauna*, and more especially to the fishes of this district, could not fail to bring to light many interesting facts. There is one kind of fish called the *palea*, belonging principally, if not exclusively, to these northern lakes, which has not had its natural history studied, or been itself scientifically described. Amongst herbivorous animals, the foremost places must be assigned to the elk and deer, and to the brown bear, and perhaps to the wild boar. Besides these in the woods, wolves, foxes, and hares are met with, and water-fowl, but *no flies or fleas!*

'The Vuig forests consist mainly of pines, firs, birches, aspens, the mountain ash, hoary-leaved alder (*Alnus incana*) elder, dwarf birch (*Betula nana*), and some dwarf willows, cherry trees, honeysuckle, the sloe, and the wild rose. Of the suffruticose bushes, which constitute what may be called an under-covering of the soil, the following are met with—blaeberries, cranberries or *Brousnika*, crowberries, stone brambles, &c., and the ground is almost everywhere covered with marsh moss (*Sphagnum palustris*), or reindeer moss (*Cladonia rangiferina*); the former on marsh or bog land, and low damp places, forming a layer of great depth; the second covering higher lying places, hills, and frequently growing on bare granite. Sometimes along with the sphagnum there grows another moss called *Kijkum*, or flax.'

Mr Judræ did not meet with any hardy or broad-leaved tree besides those named, although there is reason to believe that some grow in the northern part of an adjacent district. From the result of his inquiries at the peasants, he was satisfied that there were none here, as they seemed to have no idea of such trees as he described. 'For the first appearance of such,' says he, 'they must be looked for in the eastern, not in the northern, part of the Government of Olonetz. They have been found in the districts o Pondoj and Kargopol, lying to the east, and under a lower latitude than the forest of Vuig.

'On becoming acquainted with the distribution of different kinds of trees in these forests, we come unconsciously to the conclusion that the true primitive and aboriginal trees must have been the pines and firs. Broad-leaved trees appear as if by chance accident in places which have been burned, or which have been otherwise cleared, and having once gained a footing, they have in a time more or less protracted at length gained the mastery over the conifers. The distribution of the conifers, I have had the opportunity of observing on so many estates, under so many varying conditions, that I cannot but make this bold deduction.*

'Wherever the birch comes into collision with the pine, the latter, having a less rapid growth and comparatively limited means of reproduction, gives place to its more favourably conditioned competitor for possession of the ground. I once had an opportunity, while in charge of a forest estate in another Government further to the south, of noting the progress of this death-struggle between different kinds of trees. Over an area of 1000 deciatins, birch trees, along with a smaller proportion of aspens, have in forty years entirely superseded the pine, though within the memory of the older inhabitants their place was occupied by large pine trees, and I must add that the soil was not particularly good for the growth of birch; but in this case nature was aided by the unwise way in which the pine trees were cut down.

'On the Vuig estate the process is not perceptible, but nevertheless it is going on, though very slowly. In the neighbouring villages one sees more of the birch and aspen than in those parts of the forest which are less

---

* My difficulty is to account, in accordance with this supposition, for the presence of seeds of broad-leaved trees in the ground. In Finland I found the opinion existing that the seeds had not been lying dormant there, but had been blown thither after the conifers had been burned. But I am not satisfied that it is so.—J. C. B.

populous. This is in accordance with the view I have advanced, that in the struggle for existence these broad-leaved trees always remain master of the field, wherever man begins to destroy the primitive forest.

'The forests are virgin forests, but these virgin forests cannot be compared with the luxuriant vegetation of the tropical forests, where the favourable condition of soil calls forth a proper growth, which, unchecked, makes the equatorial forests unapproachable by anything seen here, where on the contrary, there is a poor soil and a severe climate. These make growth a slow process; and, on the other hand, storms, gales, frosts, and fires destroy an immense number of trees, which with difficulty have attained in the course of centuries considerable size. Under ruins of forests thus produced there frequently accumulates water, creating marshes, which are met with in every direction. Everywhere in traversing these forests one meets with fallen decaying trees, overturned by storms, or reduced near the ground by fires. Everywhere are these met with; and everywhere also are seen masses of stones and bare granite boulders, while peat and moss bogs cover extensive areas, in a word, it may be said, moss and lichens, bogs, stones, fallen wood—these are the characteristic features of the Northern forest. The virgin forests of the tropics are characterised by profuse and luxuriant vegetation; those of the Polar regions are the products of nature which here is poor, and often destroys her own progeny in their early or unspent life, and provides but little nourishment for them as they grow up, and accordingly we find in the forests of Vuig that the growth of trees is slow. A pine tree does not attain its exploitable maturity of ten *vershoks* in thickness at a height of ten *arschins*, or $23\frac{1}{3}$ feet from the ground in less than from 150 to 170 years, and a critical age with the trees appears to be that extending from 50 to 80 years. Trees of larger dimensions are generally trees of from 300 to 350 years' growth; but these have often such defects as unfit them for use. Defective trees, which are extensively disseminated over these forests, may have these defects attributed generally to one or other of two causes—1, External injury from fire, wind, frost, &c.; 2, Imperfect growth from the hard, stony soil, preventing the full and regular development of roots.

'The following are defects which I have frequently observed: trees broken by the weight of snow, or killed partially or completely by frost; trees with stag-horned crowns, often the consequence of fires; trees split throughout their entire length, the effect of their having been bent by the wind; trees with knots of different kinds, each of which has its local name; trees rotten or decayed in the heart. The last-mentioned defect is often manifestly traceable to injury sustained by the tap root. The defect, however, is not such as will make the forester or the trader reject the tree; very often it does not penetrate far up into the tree, and with the lower portion of the trunk sawn off, the remainder is faultless.

'From what has been stated, it may be seen that these woods do not present us with exhaustless treasures, as many rashly think. We are not merely consuming the annual produce, or a justifiable percentage on our capital; but, on the contrary, we are destroying and using up capital which nature has, with immense labour, accumulated in the course of many centuries.

'In regard to official work, I may state that this consists in selecting and designating districts for all forest operations. Under the name of Vuig forests are now included all the forests lying on the river Vuig and its affluents. The principal terms of sale are the following:

'1. They include pine wood from the forest estates of the Government of Archangel, in the district of Kem, and also from the forest estates of the Government of Olonetz, in the district of Polonetz, on the declivity towards the White Sea.

'2. The contractor is bound to prepare, in the course of fifteen years, not less than 450,000 or more than 750,000 logs, from 10 to 16 *arschins*, $23\frac{1}{3}$ to $37\frac{1}{3}$ feet long, and of the thickness of six or more vershocks.

'3. The wood so prepared may be exported, and the contractor has liberty to erect sawmills, to build ships, to navigate the same on the White Sea, and to erect his own wharfs.

'The Vuig forest being of great extent, and not having heretofore produced any revenue, it is now of great importance.'

The Russian term for a forest is *Metsa*; for a wood, *Pinta*.

The following are the names of some of the common trees in Russ English:—

English.	Russ.
Ash,	Yaseen.
Alder,	Olha.
Apple,	Yablone.
Aspen,	Oseena.
Beech,	Book.
Birch,	Bereza.
Cherry,	Beeshna.
Chestnut,	Kashtan.
Elder,	Buseena.
Elm,	Vyass.
Fir,	Hanka.
Mountain Ash,	Rabcena.
Oak,	Doob.
Pear,	Grushna.
Plane,	Yavor.
Sycamore,	Leekaya.
Willow,	Eva.

The Russian *vershok* is equal to about 1¾ inches; the *arshin* to about 28 inches; the *sajeen* = 7 feet; the *decitin* = 60 × 40, or 2,400 square *sajeens*, or 2·69972 English acres; a *verst* = two-thirds of a mile.

Section III.—Classified List of Plants.

The following is a list of plants found in the vicinity of Lake Onega, supplied to me by Forst-Meister Alexander K. Guenther, of Petrozavodsk, in charge of State Forests in the district:—

Phanerogams.

Angiosperms—Dicotyledons.

Ranunculaceae.—Atragene alpina L. Thalictrum aquilegiaefolium L. T. flavum L. T. angustifolium L. T. simplex L. Anemone nemorosa L. A. ranunculoides L. Myosurus minimus L. Ranunculus polyanthemos L. R. repens L. R. acris L. R. auricomus L. R. cassubicus L. R. lingua L. R. flammula L. R. reptans L. R. sceleratus L. Batrachium trichophyllum Chaix. B. (confervoides Fr. var?) admixtum W. Nyl. Ranunculus aquatilis, var. succulentus Koch? R. confusum Gren et Godr., R. heterophyllum Fr. R. var. peltatum Fr. R. flaccida Trtv. R. capillacea D. C. Ficaria ranunculoides Fr. Caltha palustris L. Trollius europaeus L. Aquilegia vulgaris L. Delphinium consolida L. D. elatum Aconitum septentrionale Kölle. Actaea spicata L.

Nymphaeaceae.—Nymphaea alba L. N. var. minor D. C. Nuphar luteum, Sm. N. intermedium Ledeb. N. pumilum D. C.

Papaveraceae.—Chelidonium majus L.

Fumariaceae.—Corydalis solida Sm. Fumaria officinalis L.

Cruciferae.—Nasturtium amphibium L. N. palustre D. C. Barbarea stricta Fr. Arabis sagittata D. C. Turritis glabra L. Cardamine pratensis L. C. amara L. Sisymbrium sophia L. S. thalianum Gay. Erysimum cheiranthoides L. E. hieraciifolium L. Brassica campestris L. Farsetia incana L. Draba memorosa L, Camelina sativa Fr. C. foetida Fr. Subularia aquatica L. Capsella bursa pastoris Mönch. Thlaspi arvense L. Lepidium ruderale L. *Neslia paniculata L. Bunias orientalis L. Raphanus raphanistrum L.

Droseraceae.—Drosera longifolia L. D. rotundifolia L. Parnassia palastris L.

Violaceae.—Viola palustris L. V. epipsila Ledeb. V. umbrosa Fr. V. collina Bess. V. mirabilis L. V.

sylvatica Fr.   V. arenaria D. C.   V. canina L.   V. tricolor L.   V. var. arvensis Murr.

Polygaleae.—Polygala amara L.

Silenaceae.—Dianthus superbus L.   D. deltoides L.   Gypsophila muralis L.   Silene inflata Sm.   S. nutans L.   S. tatarica Pers.   Melandrium vespertinum Fr.   Lychnis viscaria L.   L. flos-cuculi L.   Agrostemma githago L.

Alsinaceae.—Spergula arvensis L.   Lepigonum rubrum Fr.   Sagina procumbens L.   S. nodosa Fenzl.   Arenaria serpyllifolia L.   A. trinervis L.   A. lateriflora L.   Stellaria nemorum L.   S. holostea L.   S. media L.   S. glauca With.   S. graminea L.   S. longifolia Fr.   S. crassifolia Ehrh.   S. uliginosa Murr.   Malachium aquaticum Fr.   Cerastium vulgatum L.

Elatineae.—Elatine hydropiper L.   E. triandra Schk.

Tiliaceae.—Tilia parvifolia Ehrh.

Malvaceae.—Malva borealis Wallm.

Gruinales.—Geranium sylvaticum L.   G. pratense L.   G. palustre L.   G. bohemicum L.   Erodium cicutarium L.   Oxalis acetosella L.   Linum catharticum L.

Hypericineae.—Hypericum quadrangulum L.

Acerineae.—Acer platanoides L.

Balsamineae.—Impatiens noli-tangere L.

Rhamneae.—Rhamnus frangula L.

Papilionaceae.—Trifolium pratense L.   T. medium L.   T. repens L.   T. spadiceum L.   T. agrarium L.   Melilotus alba Lam.   M. officinalis Lam.   Vicia hirsuta Koch.   V. sylatica L.   V. cracca L.   V. sativa L.   V. sepium L.   V. angustifolium L.   Lathyrus pratensis L.   L. sylvestris L.   L. palustrus L.   Orobus vernus L.

Drupaceae.—Prunus padus L.

Rosaceae.—Spiraea ulmaria L.   Geum urbanum L.   G. rivale L.   Potentilla tormentilla Scop.   P. norvegica L.   P. argentea L.   P. anserina L.   Comarum palustre L.   Fragaria vesca L.   Rubus humilifolius C. Mey.   R. chamaemorus L.   R. arcticus L.   R. saxatilis L.   R. idacus L.   Alchemilla vulgaris L.   Rosa acicularis Lindl.   R. cinnamomea L.

Pomaceae.—Cotoneaster vulgaris Lindl. Sorbus aucuparia L.

Onagrarieae.—Epilobium angustifolium L. E. montanum L. E. palustre L. Circaea alpina L.

Halorhageae.—Myriophyllum spicatum L. M. alterniflorum D. C. Hippuris vulgaris L.

Callitricheae.—Callitriche verna Kütz. C. polymorpha Lönnr. C. autumnalis L.

Ceratophylleae.—Ceratophyllum demersum L.

Lythrarieae.—Lythrum salicaria L. Peplis portula L.

Portulacaceae.—Montia fontana L.

Sclerantheae.—Scleranthus annuus L.

Crassulaceae.—Sedum telephium L. S. acre L.

Grossularieae.—Ribes nigrum L. R. rubrum L.

Saxifrageae.—Saxifraga caespitosa L. S. hirculus L. S. nivalis L. Chrysosplenium alternifolium L.

Umbelliferae.—Cicuta virosa L. Aegopodium podograrium L. Carum carvi L. Pimpinella saxifraga L. Sium latifolium L. Conioselinum tataricum Fisch. Angelica sylvestris L. Peucedanum palustre Koch. Heracleum sibiricum L. Cerefolium sylvestre L, Chaerophyllum Prescotti D. C. C. aromaticum L.

Adoxeae.—Adoxa moschatellina L.

Corneae.—Cornus suecica L.

Valerianeae.—Viburnum opulus L. Valeriana officinalis L.

Caprifoliaceae.—Lonicera xylosteum L. L. coerulea L. Linnaea borealis L.

Rubiaceae.—Galium aparine L. G. triflorum Michx. G. boreale L. G. trifidum L. G. palustre L. G. ulignosum L. G. mollugo L. G. verum L.

Dipsaceae.—Trichera arvensis Schrad. Succisa pratensis Mönch.

Compositeae.—Tussilago farfara L. Petasites frigida Fr. Solidago virgaarea L. Erigeron acris L. E. Mülleri Lund. Inula salicina L. I. helenium L. Bidens tripartita L. B. cernua L. Filago montana L. Antennaria dioica L. Gnaphalium sylvaticum L. G. ulignosium L. Artemisia vulgaris L. A. absinthium L. Tanacetum vulgare L. Achillea millifolium L. A. ptarmica L. A. cartileginea Ledeb. Anthemis arvensis L. A. tinctoria L. Pyrethrum corymbosum Willd. Matricaria inodora L. Crysanthemum leucanthemum L. Senecio vulgaris L. Ligularia sibirica Cass. Cirsium lanceolatum Scop. C. palustre Scop. C. oleraceum Scop. C. heterophyllum Scop. C. arvense Scop. Carduus crispus L. Lappa tomentosa Lam. L. minor D. C. Carlina vulgaris L. Saussurea alpina D. C. Centaurea jacea L. C. phrygia L. C. scabiosa L. C. cyanus L. Lapsana communis L. Leontodon autumnalis L. L. hispidus L. Var. vulgaris Ascher, Var. hastilis L. Picris hieracioides L. Hypochaeris maculata L. Taraxacum officinale Web. Mulgedium sibiricum Less. Sonchus arvensis L. S. oleraceus L. S. asper L. Crepis tectorum L. C. biennis L. C. paludosa Mönch. C. sibirica L. Hieracium pilosello L. H. auricula L. H. Blyttii Fr. H. flammeum Fr. H. praealtum Vill. H. dubium L. H. cymosum var. pubescens Lindl. H. murorum L. H. caesium Fr. H. bifidum Kit. H. vulgatum Fr. H. praenanthoides, Vill. H. umbellatum L.

Lobeliaceae.—Lobelia Dortmanna L.

Campanulaceae.—Campanula glomerata L. C. cervicaria L. C. latifolia L. C. ranunculoides L. C. rotundifolia L. C. persicifolia L. C. patula L.

Vaccinieae.—Vaccinium uliginosum L. V. myrtillus L. V. vitis idaea L. Oxycoccus palustris Pers. O. microcarpus Turcz.

Ericineae.—Arctostaphylos uva-ursi L. Andromeda polifolia L. Cassandra calyculata Don. Calluna vulgaris Salisb. Ledum palustre L.

Pyrolaceae.—Pyrola uniflora L. P. rotundifolia L. P. chlorantha Sw. Pyrola media Sw. P. minor L. P. secunda L. Monotropa multiflora Scop.

Gentianeae.—Menyanthes trifoliata L. Gentiana amarella L.

Polemoniaceae.—Polemonium coeruleum L. P. pulchellum Bunge.

Convolvulaceae.—Convolvulus arvensis L. Cuscuta europaea L.
Boragineae.—Lycopsis arvensis L. Lithospermum arvense L. Palmonaria officinalis L. Myosotis palustris With. M. lingulata Lehm. M. Stricta Link.
Labiata.—Mentha arvensis L. Lycopus europaens L. Origanum vulgare L. Thymus serpyllum L. Calamintha acinos Clairv. Clinopodium vulgare L. Scutellaria galericulata L. Prunella vulgaris L. Glechoma herderaceum L. Dracocephalum Ruyschiana L. D. thymiflorum L. Stachys sylvatica L. S. palustris L. Leonurus cardiaca L. Galeopsis ladanum L. G. versicolor Curt. G. tetrahit L. Lamium album L. L. purpureum L. L. amplexicaule L. Ajuga reptans L.
Solaneae.—Solanum dulcamara L. Hyosciamus niger L.
Scrophularineae.—Verbascum thapsus L. V. nigrum L. Scrophularia nodosa L. Linaria vulgaris Mill. Veronica longifolia L. V. officinalis L. V. beccabunga L. V. chamaedrys L. V. Scutellata L. V. serpyllifolia L. V. arvensis L. V. verna L. Limosella aquatica L. Odontites rubra Pers. Euphrasia officinalis L. Rhinanthus major Ehrh. R. minor Ehrh. Pedicularis palustris L. P. sceptrum-carolinum Melampyrum cristatum L. M. nemorosum L. M. pratense L. M. sylvaticum L.
Lentibulariaceae.—Utricularia vulgaris L. U. intermedia Hayn. U. minor L. Pinguicula vulgaris L.
Primulaceae.—Androsace filiformis Retz. Primula officinalis L. Trientalis europaea L. Lysimachia vulgaris L. L. thyrsiflora L. L. nummularia L.
Plantagineae.—Plantago major L. P. media L. P. lanceolata L.
Polygoneae.—Rumex maritimus L. Rumex hippolapathum Fr. R. domesticus Hartm. R. acetosa L. R. acetosella L. Polygonum bistorta L. P. viviparum L. P. amphibium L. P. lapathifolium L. P. persicaria. L. P. minus Huds. P. hydropiper L. P. aviculare L. P. convolvulus L. P. dumetorum L.
Chenopodeae.—Chenopodium album L. Blitum glaucum L. B. rubrum Rchb. Atriplex patula L.

Thymeleae.—Daphne mezereum L.
Empetreae.—Empetrum nigrum L.
Euphorbiaceae.—Euphorbia helioscopia L.
Urticeae.—Urtica urens L. U. dioica L. Humulus lupulus L.
Ulmaceae.—Ulmus montana Sm. U. effusa Willd.
Salicineae.—Populus tremula L. Salix pentandra L. S. amygdalina L. S. lapponum L. S. caprea L. S. cinerea L. S. aurita L. S. depressa L. S. rosmarinifolia L. S. myrtilloides L. S. nigricans Sm. S. repens L. S. phylicifolia L. S. myrsinites L.
Betulaceae.—Betula verrucosa Ehrh. B. glutinosa Wallr. B. fruticosa Pall. B. intermedia Thom. B. nana L. Alnus incana Willd. A. pnbescens Taush. A. glutinosa Willd.

MONOCOTYLEDONES.

Hydrocharideae.—Stratiotes aloides L. Hydrocharis morsus-ranae L.
Alismaceae.—Alisma plantago L. Saggittaria sagittaefolia L. S. alpina W. Butomus umbellatus L.
Iuncagineae. — Scheuchzeria palustris L. Triglochin palustre L.
Potameae. — Potamogeton natans L. P. rufescens Schrad. P. gramineus L. P. f. helerophyllus Schreb. P. lucens L. P. perfoliatus L. P. mucronatus Schrad. P. obtusifolius M. et K. P. pusillus L. P. rutilus Wolfg. P. pectinatus L. P. marinus L.
Naiadeae.—Naias flexilis Rostk.
Lemnaceae.—Lemna trisulca L. L. polyrrhiza L. L. minor L.
Ariodeae.—Calla palustris L.
Typhaceae.—Typha angustifolia L. Sparganium ramosum Huds. S. simplex Huds. S. fluitans Fr. S. natans L. S. minimum Fr.
Orchideae.—Orchis maculata L. O. var. angustifolia Hartm. O. Traunsteineri Saut. O. var. curvifolia W.

Nyl. O. latifolia L. O. incarnata L. Gymnadenia conopsea L. Coeloglossum viride Hartm. Platanthera bifolia Rchnb. Ophrys myodes L. Epipogium aphyllum Gm. Epipactis latifolia Sw. E. atrorubens Hoffm. E. palustris Crantz. Listera ovata R. Br. L. cordata R. Neottia nidus-avis Rich. Goodyera repens R. Br. Malaxis paludosa Sw. M. monophyllos Sw. Corrallorrhiza innata R. Br. Calypso borealis Salisb. Cypripedium calceolus L.

Irideae.—Iris pseudacorus L.

Asparageae.—Paris quadrifolia L. Maianthemum bifolium D. C. Convallaria majalis L. C. polygonatum L.

Liliaceae.—Gagea minima Schulz. Allium oleraceum L. Do. schoenoprasum L.

Nartheciaceae.—Tofieldia borealis Whlnb.

Juncaceae.—Juncus conglomeratus L. J. filiformis L. J. alpinus Vill. J. compressus Iacq. J. bofonius L. J. stygius L. Luzula pilosa L. L. multiflora Lej. L. var. pallescens Whlnb.

Cyperaceae.—Schoenus ferruginens L. Rhynchospora alba Vahl. Scirpus palustris L. S. acicularis L. S. caespitosa L. S. pauciflorus Lightf. S lacustris L. S. sylvaticus L. Eriophorum alpinum L. E. vaginatum L. E. angustifolium L. Eriophorum latifolium Hoppe. E. gracile Koch. Carex dioica L. C. capitata L. C. pauciflora Lightf. C. chordorrhiza Ehh. C. teretiuscula Good. C. paradoxa Willd. C. stellulata Good. C. leporina L. C. heleonastes Ehrh. C. canescens L. C. loliacea L. C. tenella Schk. C. vulgaris Fr. C. uncella Fr. C. caespitosa L. C. stricta Good. C. acuta L. C. Buxbaumii Whlnb. C. irrigua Sm. C. limosa L. C. ericetorum Poll. C. globularis L. C. digitata L. C. sparsiflora Whnb. C. pallescens L. C. capillaris L. C. Oederi Ehrh. C. flava L. C. filiformis L. C. vesicaria L. C. ampullacea Good.

Gramineae.—Anthoxanthum ordoratum L. Hierachloa borealis Schrad. Digraphis arundinacea Trin. Alopecurus geniculatus L. A. fulvus Sm. Settaria viridis P. B.

Phleum pratense L. S. alpinum L. Milium effusum L. Agrostis stolonifera L. A. vulgaris With. A. canina L. A. spica-venti L. Calamagrostis arundinacea Roth. C. epigeios Roth. C. phragmitoides Hartm. C. lanceolata Roth. C. stricta Hartm. Phragmites communis Trin. Melica nutans L. Aira flexuosa L. A. caespitosa L. Fluminia arundinacea Fr. Enodinm coeruleum Gaud. Glyceria remota Fr. G. spectabilis M. K. Glyceria fluitans R. Br. Poa annua L. P. alpina L. P. nemoralis L. P. trivialis L. P. sudetica Haenke. P. pratensis L. Briza media L. Bromus arvensis L. B. secalinus L. Scolochloa festucacea Link Dactylis glomerata L. Festuca ovina L. F. rubra L. F. elatior L. Lolium perenne L. Elymus arenarius L Triticum repens L. T. caninum Schreb. Nardus stricta L.

GYMNOSPERMAE.

Abietineae.—Pinus Cembra L. P. sylvestris L. P. abies L. P. var. medioximae W. Nyl. Larix sibirica Ledeb.
Cupressineae.—Juniperis communis L.

CRYPTOGAMAE.

Equisetaceae.—Equisetum arvense L. E. pratense Ehrh. E. syivaticum L. E. limosum L. E. hiemale L. E. scorpioides Michx. E. variegatum Schleich.
Marsiliaceae.—Isoëtes lacustris L. I. echinospora Dur.
Lycopodiaceae.—Lycopodium selago L. L. annotinum L. L. clavatum L. L. compianatum L. Selaginella spinulosa Al. Br.
Opdioglossaceae.—Botrychium lunaria Sw. B. lanceolatum Gm. B. virginianum Sw.
Polypodiaceae.—Polypodium vulgare L. P. phegopteris L. P. Dryopteris L. P. Robertianum Hoffm. Woodsia ilvensis R. Br. P. hyperborea R. Br. Cystopteris fragilis Bernh. Polystichum thelypteris Sm. P. filix-mas Roth.

P. cristatum Sw.   Aspidium spinulosum Sw.   Asplenium filix femina Sm.   A. crenatum Fr.   A. trichomanes L.   A. viride Huds.   A. ruta-muraria L.   A. septentrionale Hoffm.   Pteris aquilina L.   Struthiopteris germanica Willd.

Mr Guenther, in a pamphlet entitled *Materialee K. Flora Obonechskago Kran*, has supplied a good deal of information in regard to the distribution and natural history of many of the plants, and he has referred to the following notices as supplying more :—

(1.) *W. Nylander*, Collectanea in Floram Karelicam p. 109.   Continuatio p. 183 (in *Notiser ur Jällskapets pro fauna et flora Fennica*, pars II., 1852.)

(2.) *J. P. Norrlin*, Om Onega-Karelens vegetation 1871.

(3.) *J. P. Norrlin*, Flora Kareliae Onegensis, par. I.

(4.) *J. P. Norrlin*, Flora Kareliae Onegensis, par. II. (in Meddelanden Societas pro fauna et flora Fennica 1876.)

(5.) *Fred Elfving*, Anteckningar om vegetationen kring floden Svir, p. 113 (in Meddelanden Societas pro fauna et flora Fennica, 1878.)

Section IV.—Vegetation in Lapland.

In Lapland, as has been mentioned, 'Wahlenberg's edition of the *Flora Lapponica* describes 1087 species of plants found in Lapland, more than double the number observed by Linnæus. Of this number only 496 are perfect plants; the remaining 591 are cryptogamous. Of grasses there are 102 species; of algæ, 55; of fungi, 94; of musci, 200; and of lichens, 207. Of the perfect plants, the snowy Alps contain 93 species; the subalpine region, 125; and the woody region, about 313. Of trees (reckoning the salices) there are 26 kinds; consisting of the Scotch fir, spruce fir, birch, alder, poplar, mountain ash, bird-cherry, and nineteen species of willows. There are no fruit trees in the country, but a variety of berries are spontaneously produced, such as black currants, raspberries, crowberries, juniper-berries, bilberries, and the Norwegian mulberry, which grows upon a creeping plant, and is greatly esteemed as an antiscorbutic.* In the gardens towards the south are raised cresses, spinach, onions, leeks, chives, orache, red cabbage, radishes, mustard; currants, barberries, elder-berry; wild-rose, columbines, rose-campions, carnations, sweet-williams; potatoes about the size of poppy-heads; French beans, broad beans, and tobacco when carefully managed; but neither white cabbage nor pease come to any perfection; and apples, pears, plums, and cherries scarcely grow at all, though cultivated with the greatest attention. The most abundant native vegetables are sorrel, which is of great service on account of its antiscorbutic properties; angelica, which is highly relished as an article of food; and the lichen rangiferinus, which furnishes the chief subsistence of the reindeer during winter, and which the Laplanders frequently boil in broth for their own use. Of the indigenous fruits, the most delicious is the berry of the rubus articus; which, when sufficiently ripened, is said to be superior in fragrance and flavour to the finest raspberries or strawberries. A small plateful fills an apartment with a more exquisite scent than the finest perfumes; and it is preserved in Sweden as one of the finest sweetmeats.'

On the *tundra*, land between the forest zone and the sea we find where the soil is pretty dry that lichens abound; on moister land these are varied with the Iceland moss, and in the southern stretches this is succeeded by grasses, cruciters, saxifrages, carophyls, and compositae and marsh plants varying the scene, but it is a dreary waste.

In Swedish Lapland there ripen rye, barley, the raspberry, the strawberry, the red gooseberry, the cowberry, and whortleberry, and the delicious Arctic bramble (*Rubus arcticus*); but neither fruit trees, wheat, or pease come to maturity.

---

* The plants on the western part of Lapland, towards the sea, are analogous to those of Scotland and Iceland; while the most abundant productions of Swedish Lapland more nearly resemble those of Siberia.

Section V.—Palaeontological Botany.

By the study of *débris* of vegetation found preserved from ancient times in peat bogs, much may be learned in regard to the state of the earth and the climate, in the place of their production, at the time of their growth; and thus may much be learned in regard to the state of the earth in these localities in times long preceding what is called the historic period, or that in regard to which we have notices, more or less explicit, preserved in historical records, and it may be long anterior to the times in which these records were made, and to earlier times to which these records may allude; and by the study of fossil plants much may in like manner be learned in regard to the state of the world in even pre-adamic times, in times much more remote from the present than the so-called tertiary and post-tertiary periods of the geologist. By the study of these we may be carried back to what may seem to be the beginning of the creation of the organic structures, vegetable and animal, with which the earth is now clothed and peopled.

Through including within the range of our observation and study the outlying islands of Nova Zembla and the lands beyond, we have been introduced to a region which, according to the rules of study in palaeontology, has been —or, if it be required of us to speak more guardedly—may have been that first site of vegetation upon the earth, from which it went forth to multiply and replenish the earth as man went forth from his first home to fulfil what was his mission in the providence of God; and the facts upon which this conclusion is based will be found well deserving of consideration.

Large and valuable collections of fossils found in the Polar regions have been made. These have been studied with much care and attention by students of fossil plants, and pre-eminently by Dr Oswald Heer, Professor of

Botany in the University of Zurich, and Director of the Botanic Garden in that city, whose decease we now deplore, and of the importance of whose work it is impossible to speak in too high terms. 'Feeble in body, bedridden for years, but indefatigable in despite of his infirmities, applying his clear vision and his extensive and varied knowledge to the pursuit of an object, the great value of which was apparent to him from the commencement of his investigations, he has become like the Pole, the covered secrets of which he has unveiled—the immobile point towards which, during the last ten years and more, have gravitated the pioneers of the North, the illustrious navigators, the skilled explorers, men of science, and men of action—when occasion called for it, men of suffering—who have traversed in all directions the Arctic solitudes, to survey their coasts, to search into their cliffs, to sound their depths, and lastly, to bring back as trophies cases of fossils and minerals, which have become the possession of the museums of Dublin, of London, of Copenhagen, and of Stockholm, but at the cost of unceasing deeds of courage.' Such are the terms in which he is spoken of by one who has followed him in his special studies.

Results of these studies have been embodied by him in a work entitled *Die Fossile Flora der Polarlaender*.

We may find our interest in a story marred by our being told by another, whilst we are engaged in reading it, what is the plot and what is the issue of it; but I believe it will be otherwise if I pause to state what are the conclusions at which Dr Heer arrived from the study of these and other fossils, and indicate the course of reasoning by which these conclusions have been attained. His conclusions, stated briefly, are, that vegetation may have first made its appearance in the vicinity of the North Pole, and thence spread southwards towards the southern hemisphere: the indications of this having been the case being, amongst others, these: remains of plants, more especially of arborescent plants and trees, similar to those of which remains have been found in the Arctic regions, have been found occurring again and again in the direction indicated, and this up to the Equator, if not beyond it. These occur in deposits of the same antiquity everywhere, and with only such modifications as can be most satisfactorily accounted for on the supposition that the first appearance of the type was in the Polar regions. The modifications are such as altered conditions of growth might induce—it being more in accordance with what is known of the laws of morphology to suppose the tropical form of the plant to be a modification of the polar one, than the polar form of the plant to be a modification of the tropical one.

With this exposition of the matter I resume my statements:—'The discoveries so happily brought to a focus by Dr Heer,' writes Count Saporta, 'have been acquired for science by the successive efforts of a multitude of travellers and at a cost of unheard of fatigue. Many of the treasures, after having been examined, or even after having been collected and carried off by force of arms, it was found necessary to abandon in whole or in part. Dr Heer cites the collections of Nierstsching in the seas about Behring's Straits, of Dr Armstrong, of Sir L. M'Clintock at Melville Island and Prince Patrick Island, and those of Dr Kane in Greenland, as having been of necessity abandoned. But others have been more happy. The American Arctic archipelago has furnished not only coal plants, collected by Sir L. M'Clintock in Melville Island and Bathurst Island, and deposited by him in the museum of Dublin; but this museum has also received from Captain Maclure cones and fossil woods from Banks' Land. The British Museum possesses fossil plants of a locality near the Polar Circle, situated on the 65th° of North latitude, near the mouth of the Mackenzie River, collected by Dr Richardson. The Alaska Territory in America, which formerly belonged to Russia, has supplied its contingent. Specimens published by Dr Heer, collected by a Finlander M. Hjalmar Turuhjelm, of Helsingfors, is only a small portion of the original collection, lost in the wreck of the vessel by which they were being conveyed—some supplied from the isle of Kugu, near Sitka or New Archangel—others from Cook's Bay on the peninsula of Alaska, 58° and 59° of North latitude. The fossil plants of Iceland have been principally collected by Professor Strenstrup of Copenhagen; they belong, like those of Alaska and of the Mackenzie River, to localities situated beyond the Polar Circle, but too close to that limit to forbid that one should seek to utilise them in a work of so comprehensive a character.

'With regard to the carboniferous flora of Bear Island, Professor Heer has received, through the medium of the Academy of St. Petersburg, rich collections of Siberian fossil plants, some come from the Island Sakhalin, at the mouth of the river Amour on the east coast of Manchooria; others are jurassic plants from the Government of Irkutsk. These are, it is true, stations situated well beyond the Polar Circle, towards the 55th° of North latitude, in nearly the same parallel as Dantzic and Copenhagen, but the ancient flora of them should contribute necessarily to clear up vividly the history of Polar vegetation, properly so called.

'The two countries in the northern region which are the most rich in fossil plants are Greenland and Spitzbergen. The ancient vegetable wealth of these centres is indicated by the numerous beds of coal which have been met with and have repeatedly been exploited, on the places which are accessible; they belong to many epochs, and consequently they mark the repetition of the same phenomena through an extent of successive ages. The characteristics of the Polar land strikes forcibly the observer who seeks to exploit them as a geologist; on the one hand the ground disappears almost everywhere as one goes to a distance from the coast, under a thick layer of ice, which limits access to the interior to a few kilometres; on the other hand the reefs, the declivities, the high beaches, and the precipitous summits of the litoral zone, wherever the action o glaciers has left them exposed and devoid of vegetable earth, show naked their uncovered skeleton, and enable us to follow with invaluable distinctness all the details of stratification and superposition, which are sometimes so difficult to verify on the continent covered with alluvial deposits, and upturned by cultivation.

'In Greenland, it is especially on the Island of Disco and along the coast stretching to the peninsula of Noursoak, that are situated the principal beds, towards the 70th° of North latitude, a little south of Upernavik, on the western coast of the region. It is there that Captain Inglefield, and Lieutenant Colomb, his second in command, on the return of their expedition in quest of Franklin, and after them Sir L. M'Clintock, and Drs Torelly and Lyell, and in the summer of 1867 M. Whymper, made successively their collections. These were submitted by their present possessors to examination by Dr Heer. But an important part of the unveiling of the Greenland plants pertains also to the Swedish scientific expedition of 1870, and to Professor Nordenskjoeld, of Stockholm, whose name is more especially associated with Spitzbergen, which was visited by him, not only in connection with the two Swedish expeditions of 1868 and 1870, but previously in 1858, 1861, and 1869, and again later, in 1872.' Of this indefatigable and successsul explorer, Count Saporta, whose statement I am quoting, wrote in 1875:—'M. Nordenskjoeld is a young savant, already famous, a true Frenchman of the North, who combines with the vivacity and sympathetic amenity of our race the spirit of thorough investigation, penetration, scientific erudition, and perseverance of purpose, in which we are too often deficient. Familiar with the nature of the North, struggling against it and subduing it, not without an effort, he has explored, at the risk of life, a land bristling with ice-peaks, almost inaccessible, but from which he has known how to bring back cargoes of minerals and of fossils. Thanks to him and to MM. Malmgren, Torrel, and others, the past of Spitzbergen is as well known to us as that of any country in Europe, it matters not what that country may be. Nothing has escaped the penetrating eye of M. Nordenskjoeld; he has found and collected thousands of specimens of fossil plants, of coals of jurassic, cretaceous, tertiary, and even of recent production, in a desolate archipelago, which is altogether devoid of roads, of means of transport, of means of access to places, and almost of the possibility of living.'

Like the greater part of the lands in the extreme north, Spitzbergen is deeply cut up; it bristles with ice-peaks, whence the name has been given. Besides the principal land, which has an outline whereby that it forms two peninsulas named respectively Eastern Spitzbergen and Western Spitzbergen, two other lands are associated with the first of these peninsulas; the first of these is North-East Land, separated from Eastern Spitzbergen by Hinlopen Straits; the other, Wiche's Land, situated to the south; while separated from this by Olga Straits, and from Spitzbergen by Stoer Fiord, is Edge Island. The entire archipelago extends over at least four degrees, from Cape South to the Seven Isles, almost touching the 81st° of latitude.

'The explorations, for which we are indebted to M. Nordenskjoeld and the Swedish expeditions, have had principally for their object the western coast. Along this coast, slashed with immense bays and deep fiords, there is no lack of fossil plants belonging to all the formations which have been specified; and from an examination of these there are obtained indications of those remote lands having formerly been continuous with Nova Zembla and Northern Russia, if not with a vast continent stretching from the Pole to the Equator and beyond it, and comprising the lands of America as well as those of Europe and Asia, and, it may be, Africa besides. The indications referred to are these:—

'As the terrestrial floras belonging to each of the formations which have been named, when observed at synchronic points, far remote from one another, manifest generally a great uniformity of aspect and composition, the conclusion is irresistible that after the emergences from the ocean of dry land upon a considerable scale, which followed the palaeozoic times, the Arctic lands, now cut up into archipelagos, must have formed part of a Polar continent of sufficient extent to allow of fresh water having been able there to play a predominating part, and of deep lakes and important rivers having become established there; and we are shut up to the conclusion that one and the same vegetation, without other divergencies than any arising from slight local diversities, occupied the whole extent of this continent, under each of the ages in which these fossils were deposited.'

At an International Congress of Students of Geographical Science, which was held in Paris in the autumn of 1875, a paper on this ancient Polar vegetation, based on the discoveries of the Swedish explorers and the work of Dr Heer, was read by Count G. de Saporta.

In this paper, referring to the views advanced by Buffon, at a time when as yet geological ideas were merely speculative without any basis of such observations as have since been made, in which he alleged that the earth in cooling must have cooled most rapidly in the Polar regions, and that lands in the extreme north 'must have enjoyed the same temperature which is enjoyed now by lands further to the south,'* he says that this is substantially correct, though facts present themselves to geologists differently from what they did to the mind of Buffon.

For the information of those to whom such studies may be altogether new, and I anticipate that such there may be amongst my readers, who may reside far from towns and libraries containing books in which they may find the information pre-supposed to be in possession of readers

---

* Buffon *Des Epoques de la Nature; Hist. Nat. Gen. et Part.* 1778. Suppl. T, ix. p. 86.

of the statements which follow, I may be allowed to introduce the following statements. According to what was advanced by Laplace as a hypothesis, but which has come to be extensively accepted as a theory, the material of which the earth consists was once floating in space in widely separated masses, amongst which was in operation what is known as the force of gravitation, under the influence of which, once and again and a thousand times told, two or more of these separate masses, reciprocally attracted, would coalesce, and if they happened to approximate each other in a line diverging in the slightest degree from that of a straight line between their centres of gravity, which would occur in the vast—inexpressibly vast—majority of cases, they would begin to rotate around each other in a curve which would most likely lead to their conjunction, when the movement would issue in a rotary motion of the composite mass. This composite mass would in like manner come under the reciprocal attraction of other masses, single or composite, until the whole, or the greater part of the whole, mass of matter within the sphere of attraction would be condensed into a large rotating mass; from this as the floating mass became further condensed, portions on the outer circumference would be thrown off by circumfugal force, as is a stone from a sling, or drops of water from a mop which is made to rotate rapidly; but these again would be gathered into smaller rotating masses, which would revolve around the central mass, which would finally be condensed as is the sun, while these smaller bodies would be condensed as are the planets, each rotating at the distance from the sun at which it was thrown off; and they, in the course of their condensation, would throw off lesser masses, condensing into satellites or moons, or into elongated masses, like the rings of Mercury.

Again, not to do more than merely mention the phenomena of volcanos, there are in many rocks appearances which have led careful observers of them to conclude that what is solid of the earth is a mere crust around a mass of molten matter; and modern science tends to show that the mere act of condensation which has been referred to, might suffice to produce a development of heat sufficient to have brought the whole mass into a state of fusion.

If this did happen, a film of crust would be formed through the cooling of this mass; and as the process of cooling advanced this crust would become thicker and thicker. By the combination of oxygen with hydrogen, whencesoever these had come, and in what way soever the combination was brought about—most probably by fire—there came into existence an immense body of water, probably at first in a state of invisible vapour, but thereafter condensed into a state of mist or cloud, and subsequently into a liquid mass constituting the ocean now covering a great extent of the earth's surface, and at places miles deep. By the movements of this, large portions of the remaining solid mass mechanically severed, or chemically decomposed, were carried about and ultimately deposited; but at first often to be again fused by the heat of the molten mass enclosed in the crust; and thus were produced the so-called secondary or transitionary rocks, the gneiss and schist, and metamorphic rocks, the last named rocks to some extent crystallised or otherwise changed by the action of fire.

The correctness of the opinion that the earth is at present a molten mass of matter enclosed in a solid crust has been called in question. But the granite or primitive rock defies all attempts to penetrate it to any thing like the depth which would enable us to determine the fact by observation and the thickness of this crust, if crust only it be; though the strata overlying this which have been deposited from water, and afterwards fused, have been so fractured and dislocated that the measurement of the thickness of some of these has been proximately determined. In these are no remains of organic structures of animal or of vegetable origin, nor could such have been expected to survive the fusing heat to which these strata have been subjected; and from the fact mentioned, however it has been brought about, they have been characterised as *azoic*, devoid of all indications of life. With the strata superimposed upon them it is otherwise, and at this point the subject is taken up by Count Saporta in the paper cited.

Above the *azoic* strata are superimposed the *Silurian*, a designation originating with Sir Roderick Murchison, derived from *Silures*, the name of an ancient tribe which inhabited a district of country between England and Wales, in which the rocks so designated are very distinctly developed, but a deposit which is very widely diffused; the *Devonian*, so named because it happens to be very extensively preserved in Devonshire, but which also is very widely diffused, and is known also as the Old Red Sandstone, in contradistinction to a later formation designated the New Red Sandstone; the mountain limestone, so called in contradistinction to cretaceous and chalk deposits of a later date; and carboniferous strata, otherwise known as the coal measures, which are generally found superimposed upon, but sometimes alternating with, deposits of the mountain limestone.

In reference to the great extent of azoic strata, gneiss, and crystalline schists, during the deposit of which the water still covered extensively the earth, Count Saporta alleges that the ocean did not then present conditions requisite for the support of animated structures even of the lowest order; that it must have been only in the sea, when reduced to a temperature which, though still high, would not coagulate albumen, that such could be expected to appear; that this appearance would occur in basins comparatively calm, suitable for the development and subsequent maintenance of such organisms; that there is nothing known at variance with the supposition of Buffon that this must have occurred first in proximity to the Pole; and, moreover, that there terrestrial vegetation first appeared, when vegetation first ceased to be exclusively aquatic, and appeared on land still immersed in vapours, and bathed by the tidal wave; and he goes on to say :—

'In these the earliest formations must have been humble

and feeble, as is the case with all beginnings. We cannot tell what the temperature which first admitted of this may have been, nor what were the first forms which presented themselves. Coincidences which are but of rare occurrence would have been necessary to give us fossil remains of that early period; nor are all organisms capable of leaving tangible remains. We may infer from what we know of existing organisms, that most probably these primitive organisms were of a soft consistency, integuments and skeletons being the result of later development of the primary types.

'The primitive strata, notwithstanding their thickness, supply us with few means of studying by these fossils the character of the few widely separated vegetables and animals inhabiting the waters during their deposit, but the absence of these is no proof that vital organisms were then non-existent. There are many remarkable geological facts which are of such a nature as to give rise to the thought that life had its first home, if not at the Pole itself, at least in the neighbourhood of it, and that once developed, it remained for a long time more active and more reproductive in the countries which border on the Polar Circle and the higher latitudes. The most ancient fossiliferous deposits, which are, at the same time, those most rich in organic remains, are found comprised within the Northern Zone. They abound, moreover, in the cold portion of that zone, from the 50th to the 60th degree of North latitude, and still further to the north. We meet, it is true, with Silurian formations in the south of Spain, and in America, at a latitude corresponding nearly to the 35th degree of North latitude; but the most celebrated localities are situated more to the north, in Bohemia, in England, in Scandinavia, and in the United States. The Laurentian system acquires its greatest development in Canada; and the palaeozoic rocks associated with crystalline masses cover a considerable portion of the Polar lands which stretch away to the north of the American lakes. It is evidently the same with the parts of the ocean which begird Baffin's Bay, and one portion at least of Greenland and of Spitzbergen. The Upper Devonian, the different stages of the carboniferous system, especially the mountain limestone, which represent the marine deposits immediately anterior to the era of the coal, are equally extensively spread over the regions bordering on the Pole. The Parry Archipelago, beyond the 76th degree of North latitude, Bathurst Island, Spitzbergen, towards the 79th degree of North latitude, and Bear Island, situated between Spitzbergen and the North Cape, under 70° 30' North latitude, supply repeated proofs of this, based on the observation of the characteristic features of each of these stages, in which nothing distinguishes either the minerological aspect or the fossils from what they are in Europe and in America thirty degrees further to the south. For a long time the professor of palaeontology has remarked that while the deposits of coal become exceptional in the direction of the south beyond the 35th degree, they show themselves continuously in the north under the highest latitudes. It must follow that the climatic conditions, or simply the geographical ones, belonging to the production of coal, which most observers agree in considering as having been formed in vast peat bogs, have not, during the carboniferous period, manifested themselves everywhere, but only in a zone, the southern limits of which can be traced approximately, whilst towards the north it must stretch itself very far, and extend probably even to the Pole.'

In the coal formations we have reached the remains of a period subsequent to the deposit of the great bulk of the mountain limestone. Dr. Heer thus, in *Flora Fossile Artica* (par. ii.), describes the vegetation of that age:—

'Towards the end of the Devonian period the dryland notably increased in the northern hemisphere; it was there an epoch of elevation from the depth of the sea. After this extension of continental land having proceeded on a vast scale, there began a new period, that of the coal formation. We have designated under the name of *Ursien* stage the first sub-division of this period; with this coincides the appearance of the most ancient terrestrial vegetation sufficiently rich to give us an idea of the appearance presented by the vegetation at this primitive epoch. This flora was perhaps in fact spread across the northern hemisphere, both in the old and in the new continent from 47° to 74° or 75° of northern latitude; and everywhere it shows the same character. Everywhere appeared the *calamites radiatus* which covered with its high columnar stems the marshy lowlands, whilst it great rhizomes penetrated everywhere into the boggy soil. Everywhere also made themselves to be seen associated with them the wonderful *Knorria*, the *Lepidodendron*, with stems ramified dichotomously, and leaves united in a compressed plume. The *Cyclostigma* which we find both in the south of Ireland and in Bear Island, are also rarely awanting in the bosom of the layers formed on an emerged soil; and these plants must have composed in part the forests under the shade of which the *Cardiopteris* and *Palaeopteris* stretched their stout fronds.

'This flora comprised already a pretty considerable number of species; and many of them showed themselves at the same time in regions so remote from one another, that their repeated presence warrants us to suspect the existence of a vast continent, stretching out both into the temperate and the arctic zone. The Russian carboniferous region prolonged itself probably to Bear Island; and the vegetation of this island would then have made an integral part of the lower carboniferous flora of Russia, of which it would mark the continuation towards the north—the proof that the Ursien stage must have been formed along the coasts of a great continent, resulting from the presence of fresh water animals, and pondal shells, and *nervopters*, which could not have lived but in a land sufficiently considerable to contain lakes and to give birth to rivers.

'What was the duration of this period? That is a point which one would fain determine! Then began a new sinking of the land; the formations found in brackish water, and formations purely marine recommenced; the carbonaceous schists and the mountain limestone again covered the ground, previously submerged with their vegetable imprints. The great extension of mountain limestone over different points in Europe and North America, and the small number of deposits of continental origin which it contains, shows to us that this lowering of the lands must have been the result of a general submergence. The Northern hemisphere must then most certainly have presented an entirely different aspect from what it had done during the Ursien stage. But then one sees renewed the same phenomenon as occurred at the beginning of the carboniferous period. We find that at the end of a subsequent reclothing of the ground, effected on a vast scale, the continental formation of culm, and subsequently that of the middle carboniferous strata, which marks the time when these kinds of deposits attained their greatest extension and their complete development. The flora, viewed as a whole, had changed but little during so long a period. Many of the dominant species remained such till even after this time, and they thus furnish a proof that in the mountain limestone epoch the land had never been entirely submerged, but that there remained always a certain continental emerged space sufficient to afford an asylum to these species of plants, so that, as soon as the culm by its emergence had presented to them a new space, they profited by this to extend themselves and propagate themselves more and more.

'We cannot question the great length of time which must have passed from the commencement of the *Ursien* stage to that of the culm; and during the long series of ages which then succeeded each other, the vital conditions of organised beings doubtless did not remain unchanging. It is a remarkable fact to establish, that, notwithstanding these changes, the species which were so numerous traversed the whole duration of this age, and penetrated beyond it, without experiencing any appreciable modification. The multiple forms which clothed the *Calamites radiatus* in Bear Island all reappeared in the more recent stage of the lower carboniferous strata—I would say in the superincumbent schists (*dachschiffer*)—of Moravia; but then this type lost itself, without our being able to cite instead of it any form which would be analogous to it in the middle carboniferous strata. And it is the same with the *Knorria*, the *Cardiopteris*, and *Palaeopteris*. These are facts which decidedly protest against the doctrine of the incessant and gradually progressive transformation of species which the partisans of that theory ought not to ignore. Their importance here is so much the greater that manifestly the plants of Bear Island had to live under other conditions of light than those of the Vosges, or of Ireland; as they have had to support a long winter night. It is indeed surprising that evergreen trees, as in all probability were the *Lepidodendron*, and the plants so amply leaved as the *Cardiopteris frondosa*, should have accommodated themselves to so prolonged a darkness; but in regard to this we must take into consideration the circumstance that the flora of Bear Island is composed almost entirely of cryptogams,* which could pass from the light more easily and for a longer time, than could phanerogams have done. Beyond this the climate of Bear Island must have been as favourable to the growth of vegetables as was that which prevailed in Ireland and in the Vosges, and that, although this island is situated twenty-six and a half degrees further north, since the species which they include are decidedly as large and as luxuriant in appearance, and that they have produced a layer of coal as thick as any found anywhere, besides at a corresponding level but in less high latitudes,† the heat was then still at this time distributed in an

---

* Two *Carpolithes*, according to Dr Heer, alone belonged to the phanerogams.
† The yellowish sandstone grit of Ireland presents only some thin beds of carbon in the immediate neighbourhood of the plants. In the Vosges, and generally in all the inferior carboniferous strata, we meet in no part with layers of coal of great magnitude. Such layers begin to show themselves only from the point of departure at the middle carboniferous strata, which have in consequence been designated as those of the period of the productive formation of coal.

equal manner over the surface of the globe, whilst from the miocene period there exists in this respect a well-marked inequality which has become still more pronounced in the present day. A comparative study of the marine fauna collected at Bear Island leads to similar results.

'The *Productus giganteus, P. striatus, P. punctatus*, and *P. hemisphericus*, which we know to have existed in the mountain limestone of this island, have been found almost everywhere in the mountain limestone, and possessed an extension equivalent to that of the *Knorria imbricata*, of the *Lepidodendron Velthrimianum*, and of the *Calamites radiatus*. Further, two moluscs of the mountain limestone of Spitzbergen, *Spirifer Keilhauii* and *Productus costatus*, have been found also in India; and another species, the *Productus Humboltii*, in South America, so that the Polar species stretched then to the tropics. The presence at this epoch of a climate not only equal, but also warm, is further proved by the banks of coral which were formed at Spitzbergen, and also by the great dimensions of the arborescent cryptograms, and by the ferns with large fronds preserved in Bear Island.' Thus far Dr Heer.

Count Saporta remarks on this :—' A picture so vivid, and so complete, beyond the interest which attaches to itself, is well fitted to suggest some reflections. M. Heer does not admit, and that with reason, that the entire earth has ever been submerged during the epoch of maritime invasion represented by the mountain limestone. He perceives the necessity there is for the supposition of one or more continents having existed, serving as an asylum for the plants driven back from the invaded portions, which then reappeared at the time of the culm, and of the coal, properly so called, the one class under the same form as before, the others represented by allied forms, though distinct. But M. Heer expresses astonishment that certain types were lost after the culm without again reappearing, although they had traversed without any modification the long period which just elapsed since the extreme base of the inferior carboniferous strata.'

And there follow some important remarks bearing upon the subject of evolution and development, after which are discussed the coal formations.

The beds of coal are often found underlaid and overlaid and intermixed with layers of schists, rocky matter capable of being split into thin divisions like slates, and so named from the Latin *schistus*. The accepted opinion is that coals are the remains of accumulated masses of woody matter, leaves, twigs, stems, and trunks of herbaceous and arborescent vegetables, in depressions which were alternately dry or nearly dry, and filled with water from which was deposited the schisty matter, the whole being subsequently submerged for ages by the sea, from which, in the course of these protracted ages, deposits far exceeding them in thickness were superimposed upon them, while subsequently they were under the pressure of these subjected to intense heat, whereby was effected a partial decomposition of them which resulted in the formation of the coal.

In view of this being a generally accepted opinion, Count Saporta goes on to say:—

'The most ancient land plants of which we have any knowledge have left their imprint on the schists which generally accompany the beds of coal. It does not follow, however, that beyond the submerged basins, or peat bogs, which supply a place for the deposit of the schists or of leafy sandstone girts rich in vegetable imprints, the land elevated above the level of the sea,—that is to say, the crystalline masses which represented the continents of the period, were devoid of vegetables. Far from that being the case, it is on the contrary shown by silicified seeds embedded in the gaps of the carboniferous age, that there existed then a forest vegetation, composed especially of prototypical conifers and different from that of which the coal beds have preserved the remains. The former occupied the interior of the land and the sloping portions of the soil which had been for a long time emerged; the latter frequented the low-lying places, and more especially the littoral depressions, where the fresh water coming from the interior accumulated, and gave rise to lagoons as vast in extent as they were shallow in depth. The Arctic lands, which did not then differ in heat or climate from those of our latitudes, produced, beyond all doubt, both of these two kinds of vegetables, of which the one is well known to us, thanks to the multitude of imprints which the coal deposits have preserved, though the other has scarcely left itself visible through the extreme rarity of *débris* capable of attesting its ancient existence.

'The carboniferous age must have been one of enormous duration, although we cannot for one moment suppose this to have equalled that of the Silurian. The Devonian period serves as a transition between them, and leads by insensible degrees from one to the other. The ocean was then immense in its area, and the emerged land, more extensive than one is at first disposed to admit, composed only primitive crystalline regions. Without being strongly marked in profile, or offering an *ossature* established on a very large scale, these palaeozoic lands had, however, a certain elevated contour; and the coast-line must have been marked out with some measure of distinctness. It is to emersions produced from many reprisals in such a way as to draw each time from the waters a low girdle around the continents of the epoch, that are due in reality the formations of coal, and the deposits in which these are found. Elsewhere it is always along the shore-line, and most frequently on the marine formations immediately anterior to their production, that these coal basins have established themselves. And in this respect we see well, by the descriptions of Dr Heer, and by the indications given by the celebrated Swedish explorer, Nordenskjoeld, that the Arctic localities differ in nothing, so far as their conditions can be determined, from those which have been observed on Europe belonging to the same period. The most remote in time of these emersions following deposits of coal and of carbonaceous schists, with vegetable imprints, occur towards the Upper Devonian, and it is there that are found the most ancient land plants of which we have any knowledge ; but that is not to say that these were really the first. So far from that it is in fact easy to establish that the vegetation, already far removed from the point of original departure, contained nearly the same elements as that of the carboniferous land, properly so called, save for the variations and partial modification to which the flora continued to be subjected in passing through this protracted period. The Devonian plants are rare everywhere ; and they have not yet been met with in the Arctic regions ; but in the upper portion of the Devonian between this formation and that of the mountain limestone, with its characteristic *Productus* and *Spirifer*, there is seen on a pretty great number of points both in Europe and in the Polar Zone, a primitive coal-bed with terrestrial plants, which testifies everywhere to a great uniformity of vegetation. It is to this lower coal-bed that M. Schimper has recently applied the name of *Paleanthracitic* stage, and M. Heer that of the *Ursien* stage, so naming it from Bear Island, *L'Ile des Ours*, where it appears more developed than elsewhere. This, moreover, is embedded between two marine deposits, which proves that the sea had retired during the deposit of the carbonaceous beds which enclose the imprints it contains, and then returned to cover again the deposit after it had been formed, a deposit consequently littoral, as well as one certainly made under fresh water. The distinctive plants of this *Ursien* layer reappear not only in the Parry Islands and in Spitzbergen, but at a greater distance from Bear Island, in Iceland, near Aix-la-Chapelle, and in the Vosges, where they have furnished Professor Schimper material for an important memoir on the flora of the transition land of the Vosges.

'It happens, then, that not from a mere local accident, but from a vegetable period long anterior to that of the coals, and coincident with a series of simultaneous emersions elsewhere, the result has been obtained of making us acquainted with the principal forms which then predominated among vegetables, but only within the perimetre of a littoral zone of rather limited extent.'

The *productus* and *spirifer* spoken of are bivalves like the cockle and mussel, the shells of which are found with many others in the mountain limestone, and in it alone. What is so called is, as has been stated, a series of limestone strata lying immediately below the coal measures, and, in some cases, alternating with them. They extend over great part of Central and Northern Europe; they are found again in the lake district of America, and they extend to the borders at least of the Arctic Ocean, extending between the parallels of 60° and 70°, stretching towards the mouth of the Mackenzie River. The nature of the organic remains found in them, as well as the continuity of the calcareous beds of homogeneous mineral composition and the great thickness of the deposits, concur to prove that the whole series was formed in a deep and extensive ocean, in the midst of which, however, there were many islands. Amongst other characteristic fossils are the *encrinites*, popularly known in some localities as St. Cuthbert's beads and ammonites, and the bivalves mentioned.

After tracing the relations of numerous allied plants, and the characteristics of those found in different localties, and the successive changes observable in strata of successive formation, he states that the primæval type, or palæozoic stock, of the *Salisburias* and their allies appears to be the *Psygmophyllum* of Schemper, and that in all the circumstances of the case it might have been expected that some remains of plants possessing the same characteristics would be found in Bear Island. Such he considered remains figured by M. Heer under the name of *Cardiopteris polymorpha et frondosa.**

Count Saporta then gives some details in regard to the

---

* *Kohlen fl. d. Bären.* Insel.; tab. xlv., fig. 1-4.

geography and geognosy of the Polar region in regard to the explorers and the different beds or deposits found there; and, resuming his palæontological narrative, he says:—

'It appears to us to be indubitable from the studies of the Arctic flora by M. Heer, that at the time of the coal formations no influence of nature acting on the climate, and through it on the vegetation can be attributable to latitude, the effects of which, difficult to determine at this distance of time from the events, are found to have been entirely neutralised, if not annulled.

'We have no facts obtained from the Permian, of which we have not indications all the way up to the Pole.

'The Trias exists at Cape Thordsen in the basin of Isfiord, where M. Nordenskjöld has collected not only the plants of this deposit, but marine fossils characteristic of it, and amongst them remains of the *Enalosauriens*, of the genus *Ichthyosaurus*, the presence of which testifies that the great swimming reptiles, then so diffused in the seas of Europe, were not excluded from the circum-polar seas. This is an important indication of the equality of climate; and this climatic equality among the terrestrial zones is further established by an examination of the Jurassic vegetables of Cape Boheman.' These, though not yet published at the time Count Saporta's paper was read, had been described by Professor Heer, and drawings of them had been sent to the Count, who goes on to say:—

'An immense interval of time has elapsed: since the time when the plants of Bear Island lived; the vegetation is entirely renewed; it has completely changed its aspect. The species, the genera, to some extent even the families, are no longer the same; but the changes have evidently been brought about conformably to what was going on in Europe in the same direction, and by the same process of evolution. As in Europe, the vegetation, taking its departure from the same point, has led by degrees to the same results, and presents at the time at which we find it again the same characteristics as in the heart of our Continent and in France. There are the ferns with a foliage often meagre and leathery, equisetaceæ of the genera *equisetum* and *phyllotheca*, cycads, conifers, and lastly some rare monocotyledons. There are there also the forms which are the most generally diffused in England and France, in the Bathonian, the Oxfordian, and the Coralline, and which is seen at Scarborough in Yorkshire, at Mamers in France, and at St. Michael, near Verdun. The resemblance to those found at Scarborough and with those found at St. Michael is very striking.

'The flora of Cape Boheman contains thirty-two species; a third, about ten of them, have been found, moreover, and always in the inferior oolite, or the brown jura of the Germans. Amongst the ferns, the *Scleropteris pomelii* Sap., a species of St. Michael, is especially characteristic, as it denotes the presence of a genus which is essentially oolitic. The genus *Phyllotheca*, a type of equisetaceæ which has been for a long time extinct, has been found, first in Australia, and thereafter, by Professor Zigno, in the Oxford strata of the Venetian Alps. It was then there, a genus of which the extension was immense. And the Polar species *Phyllotheca lateralis*, described by Phillips and Lindley under the name of *Equisitum laterale*, has just been found in Siberia. We have here, then, a most curious type, the sheaths of which, split in segments, distinguishes it from the true *Equisitum*, and approximates to the schizoneuva of the Trias, which is found towards the centre of the jura spread over the whole land, though everywhere pretty rare.

'This was doubtless a type in a state of rapid decadence, but one which, from that circumstance, seems well fitted to throw light on the equality of temperature which was still general at that moment from one end of the globe to the other. The cycads alone number eight species out of the thirty-two, more than a fourth part of the whole; and in regard to frequency of occurrence they hold the first rank. The genus *Podozamites* predominated amongst them: this genus distantly recalls the *Zamia* of the present day, and better still the *Ceratozamia* of Mexico, but it presented more modest proportions than the last. The fructiferous cones of these plants, which have just been discovered, it appears, by a *savant* of Stockholm, companion of M. Nordenskjöld, M. Nathorst, confirm this relationship; they remind one, moreover, of the first of the two genera of the present, which have been mentioned. In Europe the *Podozamites* are often frequent at the base of the lias in the rhetien; but they reappear in the oolite, and even further up in the wealden. One of the most characteristic species of these in the deposit at Scarborough, the *Podozamites lanceolatus*, Lindl., constituted also a part of the flora of Cape Boheman.

'Other forms of the Bathonian deposit of Scarborough show themselves almost as abundantly as the preceding, at Cape Bohemau; these are the *Cyclopteris Huttoni*, Sternb., and *C. digitata*, Brongn., the place in the classified list and the peculiarities of which cannot be passed over in silence. Long considered as ferns analogous to the *Schizoca*, or by others as *rhizocarps* of a lost type, the *Cyclopteris* and the *Baiera* of Schimper, have been recognised quite lately, and with perfect justice, by M. Heer, as representing in reality the *Salisburia* (*Ginko* L.), being in reality, notwithstanding their antiquity, congeners of the unique species of the present day, *Salisburia adiantifolia*[1] Sm. (*Ginco biloba* L.)'

The *Salisburia adiantifolia*, or 'maiden-haired Salisburnia'—so named after a distinguished modern botanist—is a native of Japan, but now common in Europe. It is a tree of great beauty, attaining a height of about twenty feet. It is remarkable for its fan-shaped leaves, cloven like some of the species of *adianthum*, from which circumstance it has received its specific designation. It belongs to the same order as the yew, which order is intermediate between that of the joint-firs, the gnetaceæ, and the pines. While resembling in some points the ferns, the fruit, like that of the yew, is juicy, and resembles a berry or rather a damson, which it also resembles in size. It is of a pale brown colour, but becomes yellow when ripe. The pulp is white and fleshy, adhering closely to the drupe, which is like that of the apricot. It is in taste both resinous and astringent, and is exposed for sale in the markets in China and Japan. The kernel is white, rather firm, and sweet, with a mixture of austerity or bitterness when raw, but agreeable when roasted, and is thought by the Japanese to promote digestion.

Count Saporta considers that the primæval type, or original palæozoic shoot of the entire group of *Salisburias*, was the genus *Psygmophyllum* of Schimper, and he adds: ' If this opinion shall be confirmed, the point of departure of the *Salisburias* will fall to be placed in the palæanthractic vegetation of the extreme north. An immense interval or gap, comprising the Permian, the Trias, and the Lower Jurassic formations, forbids that we should say anything of the *Ginkophyllum*, the *Trichopitys*, the *Chiropteris*, and the *Jeanpaulia*, which represent in Europe, in these different stages, the successive forms of this group of primitive *Salisburias*. But we find it again on coming to the Arctic Jurassic flora of Cape Boheman.' . . .

'Besides the mountain limestone underlying the coal formations, we have mountains of chalk, from the littoral cliffs of which England, it is said, got its name of Albion. This is a much later formation than the mountain limestone, and the organic remains preserved in the strata show a decided advance upon pre-existing races of animals. We find in it *Zoophytes* more like existing species than were those of the mountain limestone and silurian rocks; star fishes and sea urchins resembling those of the present day; *Annulosa* like the common *Serpula* and land-worm; *Crustacea* resembling the lobster tribe; insects like the beetle and dragon-fly; fishes belonging to the *Ganoidia*, of which the sturgeon and the bony pike of the North American lakes are representative in the present day; reptiles allied to the tortoise and to the crocodile, though differing from these in external form; and two or three small mammalia allied to the opossums. Of plants, besides some previously found, and allies of these, we find plants allied to the *cycus revoluta*—the sago-plant—and the pine-apple; conifers resembling the pine, as well as yew-like and lily-like plants; and other undescribed genera. In this system of strata the arenaceous are no longer sandstones, but loose unsolidified sand; the argilaceous beds are generally soft and marly clays; and the calcareous, instead of being compact or crystaline limestones, present that soft earthy texture which prevails in chalk. All this speaks a comparatively recent formation, with freedom from great pressure, long-continued chemical actions, or the indurating effects of heat. The lower cretaceous strata found in the north, according to Professor Heer, are elevated by but a very slight degree above the Wealden. It has been met with by M. Nordenskjoeld in a series of elevations along the northern coast of the peninsula of Noursoak, at 70° 37' 43" of North latitude. These are the black schists and the grits, which alternate with each other a great many times, and repose directly on the gneiss. The total thickness of the formation attains to 1500 feet without changing sensibly in character, and the summit is found covered with overflowings of basalt. The vegetable imprints abound chiefly in the schistose beds, and more ordinarily, but not exclusively, towards the base of the formation, which it appears should be carried back in its entirety to only one and the same period. The principal collections are at Kome, at Pattorfik, at Avkrusak, at Karsok, and at Ekkorfat. The localities in the order of their richness are Kome, Avkrusak, Ekkorfat, and last Pattorfik. These localities present each of them a peculiar character: Kome abounds in ferns, but repeatedly occurring vestiges let us see between them, quite near, a forest of fir trees. Pattorfik has quite the aspect of a wood of *Sequoia*, tapestried with ferns. Ekkorfat comprises especially *Cycads*, associated with *Sequoias* and with fir trees, the combination of which created a great forest. All these localities, evidently contemporaneous, have furnished together sixty-five species, a considerable number, superior to that of the greater number of the European local floras of the same period. Nothing can be more curious to examine closely than this collection of forms then reassembled in the bosom of the same country in the neighbourhood of the Pole.

'Time has passed since the Jurassic; and it has put its impress on this new flora, and has led to many changes from the anterior state of it; but as changes have an importance almost always proportional to the time passed, and as the interval which stretches from the Bathonian, probably the level of Cape Bohemian, is infinitely less than that which separates the plants of this last from those of the lower carboniferous strata, it is quite a simple matter to establish the less profound modifications in the nature at least of the constituent elements of the Arctic vegetation in looking at it towards the commencement of the chalk period. Ferns, cycads, and conifers, compose always the principal groups; the ferns dominate in their entirety, the conifers come next. The cycads hold only the third rank in number, as well as in frequency of occurrence.

'But we have established in this respect very sensible local differences: the cycads scarcely show themselves, excepting at Kome and Ekkorfat, and always associated with ferns and with conifers; whilst at Pattorfik there are only ferns and conifers, and at Avkrusak these two groups have only by the side of them some remains of cycads; the monocotyledons do not show themselves but in a restricted number, and they have nothing conclusive about them; there were no palms as yet, as there were in Europe at the same age, but probably screw-pines rather ill-defined as yet—in fine, with scarcely a single exception, no dicotyledons; and that exception was special to Pattorfik, where the beds with vegetable imprints occupied the extreme base of the formation, which is all the more curious. It constitutes of itself an occurrence of which I shall, in a little, enquire into the exact significance.'

He proceeds to specify coincidences, such as the contemporaneous appearance in Greenland and in the Carpathian Mountains, separating Bohemia and Hungary, of the same characteristic types of cycas-like plants. Older forms of ferns seemed to die out; but of those which now appeared their representatives must be sought in the present day in the vicinity of the tropics, or in the warmer parts of the temperate zones of the northern and southern hemispheres. Thirteen species of *Gleichenia* have been identified by Professor Heer as similar to those growing at the same time in other parts of Europe. The *Gleichenias* of the present day are diffused over the tropics and the islands of the South Sea. But one species, bearing fronds once or repeatedly divided dichotomously, and generally provided with a bud situated between the branches of the dichotomous division, has advanced so far north as Japan. And the Polar *Gleichenias*, which present a physiognomy absolutely the same, must have sought the same conditions of warmth and humidity as their congeners of the present day.

'The conifers of this epoch divide themselves naturally under several categories, having each its peculiar signification. It is beyond contradiction the most important group of the period now under consideration; and amongst the types comprised in it may be remarked many which show themselves for the first time, and of which it seems that the cradle should decidedly be placed in the interior of the Arctic zone. It is there that these types, after having remained a long time confined to it, and after having there given rise to a certain number of forms, went out to spread themselves further towards the south in distinct rays, some later, in such a way as to gain access to the two continents, and there to make good their footing, and maintain it, a long time after they had disappeared from the land of their origin.

'We have spoken of the *Salisburias*—they continue to show themselves always divided into two groups, that of the *Ginko* properly so called (*Salisburia Arctica* and *S. grandis*), and that of the *Baiera* and *Jeanpaulia* with lanceolate leaves with narrow segments, represented here by the *Sclerophyllina cretosa*, Schenk., and *S. dichotoma*, Heer. But alongside the *Salisburias* appears for the first time a veritable *Taxad*—the *Torreya Dicksoniana*, Heer—a remarkable species precisely determined, which proves that the group of *Taxineas* proper had its cradle in the north, and that, after having dwelt a long time there, it passed thence into Europe, into America, and into Asia. Europe does not possess, it is true, the genus *Torreya*, but this genus has certainly lived there aforetime; and, in concert with Professor Marion, I have lately determined it in the pliocene tuffas of Meximieux under a form which it is difficult to separate from the *T. nucifera*, Sieb. and Zucc., of Japan. The *Glyptostrobus* and the *Sequoia* have followed a course in every respect alike. The *Glyptostrobus Groenlandicus*, Heer, is indeed the direct ancestor of *G. Ungeri*, Heer, and *G. Europaeus*, Brongn., which abounded in the Arctic zone in the time of the Lower Miocene; these two sister forms—forms slightly modified from the same type, spread themselves in Europe, and without doubt throughout the whole temperate zone in the course of the Miocene. Subsequently they disappeared from our continent, where, however, the *G. Europaeus* still lived towards the middle of the pliocene times. But to-day Southern China possesses, under the name of *G. heterophyllus*, a descendant scarcely modified from the *G. Ungeri* of the tertiary period.

'The chalk is veritably the age of the *Sequoia*. The *S. Reichenbachii*, Gein., obtained then an immense extension; it is found everywhere in Europe in the middle chalk and in the superior chalk. It approaches, as does the *S. gracilis*, the *S. gigantea*, which is met with in the tertiary. But by the side of this *Sequoia* there may be distinguished yet others, and amongst them the *S. Smithiana*, which, with the help of an intermediate series, connects itself without any gap to the *S. sempervirens* of California. It is then really in the north that we find the cradle of this genus also. There, after their birth, the *Sequoias* multiplied themselves, and after a first diffusion of their cretaceous species, the tertiary polar flora shows us, under forms differing little from the preceding, that these spread themselves in their turn, and invaded the whole northern hemisphere, until the time when there began to be a definite decline of the group. It is known that in our days there do not exist any other indigenous *Sequoia* than those of California, represented by two species reduced to a most restricted area of habitation, the last vestiges of a long train of forms and of subtypes.'

Similar details are given in regard to representatives of the several orders of pines, firs, cypresses, and poplars found there in the lower cretaceous deposits; and in a similar way are treated remains found in the upper deposits, including a genus *Dewalquea*, which represents in a prototypic state the *Hellebores* of the present. And having given details of specimens found in Bohemia and found in Dacota and Kansas, he goes on to say:—

'If we transport ourselves into Greenland at the epoch of the upper chalk, and look upon the collection of dicotyledons, passing over the more uncertain forms, we see that this large class comprises everywhere the poplars with leathery leaves, *Populus Bergreni* and *P. Hyperborea*, Heer; the *Ficus*, the fruit of which has been recognised, and the leaves of which were thick; the galeworts, the magnoliads, the *Credneria*, the arales, the diospyrads, the myrtaseas (*myrtophyllum*), and lastly, the leguminosae. One sees also that certain families inevitably reappear at this epoch. Let one place himself in Bohemia, in Kansas, or in Greenland, &c., and the effects of latitude, so far as they make themselves appreciable, find themselves still restricted within the narrowest limits. The frequency of poplars, the absence of laurineae with persistent leaves, and the presence of one laurineae with caducous leaves, still doubtful, it is true (*Sassafras arctica*, Heer)—such are the only indices upon which one can rely in admitting the influence of a climate already colder in the extreme north, in Europe, or in America, since the upper chalk. But the diffusion of the *Magnolia*, then present everywhere, the abundance of plane-trees, and the presence of a beech tree in America, would seem rather to favour the supposition of a very great equality in what are precisely those the floral parts of which have experienced least of reductions and adhesions of parts; in them the primitive axis, the contraction of which has given birth to the floral formation, is still recognisable, and the phyllotaxis, or order of arrangement of the accessory elements of this axis is still perceptible, at least, partially in the spiral disposition affected by the sexual organs, and even by the modified leaves which surround them. The greater part of dicotyledons, not the first doubtless, but at least the more remote from the point of original departure, have stipules; and the sheathing petiole of the arales, the long-prolonged limb of the petiole of the *Credneria*, and the frequency of the palminerved arrangement, or a tendency towards this arrangement, are in our eyes so many indices of an anterior state of the foliaceous organs towards which the phylloid floral formations of certain types are perhaps only a partial recurrence; so that the stipules appear to constitute a last vestige of these. It is then probable that the dicotyledons, at the time when we encounter them for the first time, had already been subjected to a long series of modifications. Many of them have taken upon themselves in the course of this progression, abortions and adhesions, secondary variations, the effacement of certain characteristics and conditions of climate to the north, as to the south, of the Polar Circle.

‘It is certain that the immense extension which certain forms obtained at this epoch, such as the *Sequoia Reichenbachii* and the *Gleichenia*, militate in favour of a like equalisation of temperature extending from one extremity of our hemisphere to the other, and there, perhaps, lies the whole secret of the rapid development and the general extension of the dicotyledons. The region in which the plants of this class had their first cradle, without being situated in the immediate vicinity of the Pole, could touch it, however, and communicate at the same time with the zones further south. Some day, it is to be hoped, we shall be able to fix the geographical location and probable limits of this mother region of the first dicotyledon; at present the data are too vague to allow us to think of insisting further on this point. I have been desirous, however, of investigating whether the families of the most ancient dicotyledons, and those the presence of which in the chalk age have been determined in the manner the least doubtful, present in themselves any character which would prove their antiquity. In regard to this the frequency and the diffusion of the polycarpic plants, magnolaceae, menispermeae, perhaps berberidaceae, heliboreae, nympheaceae and malvaceae, have not passed unobserved, as the excessive development of certain parts in many of these families have not ceased to reproduce and multiply the types and the subtypes within each of the species. This elaboration has gone on across the last part of the chalk period and throughout the whole of the tertiary, and it continues still in the heart of the polymorphic and floating groups which drive botanists to despair.'

Thus are we brought to the close of one of the great divisions of geologic eras.

The rocks in which are found the fossil remains which have been under consideration latterly have been called secondary in contradistinction to the granite, gneiss, and other underlying rocks which have been designated primary rocks. But there are rocks of later formation, which it has been found convenient to distinguish from those again; and to these the designation 'tertiary' has been given.

The fossils found in the older rocks presented little analogy, often no resemblance, to existing plants and animals; here, however, the similitude is frequently so complete, that the naturalist can scarcely point out a distinction between them and living races. Of land plants there have been found cycas-like plants, coniferæ, palms, willows, elms, and other species, exhibiting the true dicotyledonous structure. Nuts allied to those of the cocoa and other palms have been found in the London clay, which belongs to this class of strata; and seeds of the fresh-water characeae or stoneworts, known by the name of gyrgonites—so named from the Greek *Gyros*, curved, and *gonos*, seed, which is descriptive of their form - are found in the same deposit. Lyell has subdivided the tertiary strata into four groups: named respectively the Eocene, found in Paris, London, and Belgium—3·12 per cent. of the fossils found in which are of recent species. The name is composed of the Greek words *Eos*, the dawn, and *kainos*, recent, and was given in allusion to recent species beginning to appear; but it should be noted that the reference is to animals, and only incidentally to plants. The Meiocene, from *Meion*, less—a designation given, I presume, in relation to those which follow. This is found in Vienna, Bordeaux, Turin, &c., and contains amongst its fossils 18 per cent. of recent species. The Pleiocene, more recent, from *Pleion*, more; it is found in Italian and crag deposits, and of its fossils 41 per cent. are of recent species. And the Pleistocene, the most recent from *Pleiston*, most; this is found in Sicilian deposits, and 95 per cent. of its fossils are recent. The nomenclature proceeds on the assumption that the greater the proportion of fossils found of species which still exist in a living state, the nearer to our times must have been the period of its deposit.

The apparent uniformity of heat over the surface of the earth in earlier times, from the equator to the pole, may be attributed to the temperature of the cooling mass being so far in excess of any heat communicated by radiation from the sun to it that this scarcely disturbed the equality of the temperature anywhere. But subsequently it was otherwise, and in regard to the tertiary period, it is stated in the paper cited:—

'The chronological order we have followed has brought

us now into the tertiary formations. The Arctic flora of this epoch, which saw the Polar lands gradually cool down, become covered with ice, and finally extirpate all fruit-bearing vegetation, is the most rich in record of all those of which Professor Heer has published the plants found. We are far from being astonished at this profusion. We must bear in mind that in the north, as well as on the flank of mountains, no type of plants is represented in any degree by the most beautiful or magnificent individuals in approaching the limit marking the point of definitive arrest. The beech in Denmark, the pedunculated oak in the neighbourhood of Stockholm, the white birch in Dalecarlia and on to Altenfiord, the fir of the Alps, the pine of Norway, all supply striking proofs of this truth. It was the same of old in the Polar regions, where the ancient vegetation, after having been subjected from period to period, as everywhere else, to a gradual progress, after having acquired new types, and lost previous types, or seen the aspect of them changed and more or less modified, reached at length an age in which the heat began to decrease, in which the seasons began to show clearly their differences, and the hibernal night made them feel the effect of its long darkness. This age evidently coincides with the tertiary age; but before leaving the field to the ice masses and giving up the extreme north to devastation and solitude, the Arctic climate passed through many phases.

'We have seen that towards the close of the chalk period the reduction of temperature was as yet but little felt, but difference of latitude tended to show itself, and to do so ever more strongly.

'We have no evidence that the palms and the dodder-laurels with persistent leaves, the hibernal flowering of which required the presence of light in the cold season, have ever had their habitation within the Polar Circle.

'From the eocene to the epoch in which these plants spread themselves in Europe, and advanced at least to the 55° of latitude, the Arctic regions regions presented doubtless already winters too marked, and summers of to warmth and too short, to open up for them access Polar zone.

'The contrast between the two seasons and the d of that of winter ought necessarily, through the in of an annual period of enforced repose on vegeta favour the development of species with caducous Indeed, we are not far from admitting that the part of the types of dicotyledons with caducous must have come originally out from the e north, and that the cradle of some of them ough placed in the interior of the Arctic zone, though be that for some others of them it must be pla mountains and in the moist parts of the temperat It has been certainly thus with groups which cc at once species with caducous leaves and othe persistent or semi-persistent leaves, such as the e which the sub-genus *Microptelea* represents the ty] non-caducous leaves; the birches, of which the *B* betokens the southern stock; the oaks, divided int green oaks and common oaks; and the chesnuts, o the *Castanopsis* and the *Pasiana* are the repetition heart of the temperate zone. Every time that we can a duality of this sort, we are certain to meet in the tertiary vegetation remains of the sub-type with ca leaves, whilst the other sub-type is alacking, and itself at the same epoch by preference in Europe. types, as those of the ginko, of the plane tree, of t tree, &c., the prototypes of which, with persistent have disappeared very long ago or are unknow1 really come from the Polar region at a definite 1 spread themselves then step by step across the n temperate zone. These kinds of trees, like the pre had rayed out from the Arctic land, and their diffusion finds the occasion of its being in this a emigration, by means of which they became free to a towards the south in one or in many direction *Liquidambar*, the *Betula alba*, the *Fagus sylvatica*, th

There are supplied details relative to indications of the times at which different trees mentioned in this table and allies of these had made their appearance at different places, and it is added that amongst Arctic species may be mentioned the following as having been the stock from which have sprung the European and American plants of the present day, the names of which are associated with them in the following list:—

Artic Miocene Forms.	Derived Forms of the Present Time.
Potamogeton Nordenskjöld, Hr...	Potamogeton natans L.—Europe.
Quercus groënlandica Hr............	Quercus prinus L.—America.
Ulmus Braunii Hr.....................	Ulmus campestris L.—Europe.
Menyanthes arctica Hr..............	Menyanthes trifoliata L.—Europe.
Viburnum Whymperi Hr...........	Viburnum lantana L.—Europe.
Hedera Mac-Cluri Hr.	Hedera helix L. Var. hibernis.—Europe.
Acer otopteryx Gp...................	Acer dasycarpum Michx.—America.
Juglans acuminata Al. Br.....	Juglans regia L.—Europe.
Sorbus grandifolia Hr................	Sorbus aria L.—Europe.
Prunus staratschini Hr..............	Prunus spinosa L.—Europe.
Crataegus oxyacanthoides Gp......	Crataegus oxyacantha L.—Europe.

After some other remarks in elucidation of what has been advanced, Count Saporta goes on to say:—' We must go about 30 degrees of latitude towards the south to find, growing wild and associated in an analogous condition, the vegetable forms which flourished then in the Polar zone. The lands of that zone, at the time of which a portion of their secret has now been revealed to us, formed then a vast expanse, perhaps the only continent of the time. These lands were at the same time affected by the interior fire, exposed to incessant eruptions, and subjected to the overflow of Basalt. It is known that such a state of things is no obstacle in the way to the advancement of vegetation, and that it even favours in some cases its development in spite of the partial devastations which it brings in its train. Auvergne and the Cantal in France have been equally with this region, in a later age, the theatre of the same phenomena; and the abundance of imprints left on the ashes and volcanic mud attest that the forests must have been rich, varied, and vigorous in their growth up to the immediate vicinity of the ancient craters. It is not then in eruptive phenomena, whatever violence we may attribute to them, that we must seek for the true cause of the disappearance of the Arctic tertiary flora. This was brought about exclusively by the climate, the fall of temperature, at first almost nothing, then, scarcely perceptible in the cretaceous period, gradually increased, and from the time that it passed a certain limit it brought on necessarily the retreat or the definite disappearance of a multitude of species, which, up to that time, had been the ornaments of the lands of the north. In proportion, as this eliminatory process progressed, the glaciers, the extension of which in Europe towards the close of the tertiary period must have increased and descended from the high summits, and finally overrun all and effaced all. This invasion of the glaciers of the north, an invasion not partial as in Europe but almost universal or general, has been doubtless the proximate and direct cause of the elimination of the later tertiary vegetation; or rather it should be said, the reduction of temperature at once cause and effect, in promoting the extension of the glaciers into countries evidently very moist, has contributed by this very extension to intensify the cold, and ultimately to transform the climate, so as to render it incapable of promoting the vegetation of the greater part of plants, while by a kind of exappropriation the ground stole away and came under their denomination.'

There are embodied in these citations deliverances on two perfectly distinct questions, in regard to which a difference of opinion on either of which may exist without affecting the conclusion to which one may have come upon the other: the one relates to the course followed in the diffusion of the plants specified; the other relates to the cause or occasion of changes of temperature of which this diffusion may be considered an indication. It is the former alone with which I am conversant, and which I desire more especially to bring under the consideration of my readers. The views of Hutton and of Count Saporta, which I have cited in connection with my reference to the nebular hypothesis of Laplace, I have adduced only as a working hypothesis sufficient to remove objections which might suggest themselves to my readers, and to make comprehensible the views advanced in regard to the diffusion of vegetation from the polar region towards the equator.

To make this subject more intelligible, I may further remark: that the existing distribution of vegetables on the earth's surface has been greatly determined by conditions of soil and climate favouring or arresting the growth of plants produced by seeds dispersed from some parent plant. Amongst conditions of soil operating thus may be reckoned its constituents, including moisture and the state of disentigration in which it exists. Amongst conditions of climate may be reckoned its humidity and temperature, and the maximum, minimum, or medium measures of this. As a result of this we find that there are zones of latitude and of altitude marked by characteristic vegetation, and that there are forms of vegetation which have become characteristic of various localities; we have the palm form in some, that of the minosa in others, the coniferous form in others, the eucalyptous form in others, and the heath form in others. Professor J. H. Balfour, in a chapter on Geographical Botany in his volume entitled *Outlines of Botany*, says:—

'We sometimes meet with marked centres, where the maxima of the genus of an order, or of the species of a genus occur, the number of the genera or species diminishing as we recede from these centres, and ending perhaps in a solitary representative in some distant country. Gentians and Saxifrages have their maxima in the European Alps; Erocaulons have their great centre in Brazil, but a few species are found in other countries. Epacridaceæ are restricted to Australia. The genus Viola has two marked centres, one in Europe and another in America. The form of the European and American species are quite distinct.

The maximum of the genus Erica is at the Cape of Good Hope; but members of the Heath family extend to northern regions in the form of Erica Tetralix, E. cinerea, and Calluna vulgaris. The tropical Myrtaceæ have Myrtus communis to represent them in Europe, Leptospermeæ in Australia, and Metrosideros lucida in Lord Auckland's Group, lat. $50\tfrac{1}{2}°$ S.

'An order, or a genus, or a species, in one country is occasionally represented in another by forms which are either allied, or have physiognomic resemblance. There is thus sometimes a repetition of resembling or almost similar forms in countries separated by seas or extensive tracts of land. The Ericaceæ of the Cape have in Australia a representative in the nearly allied Epacridaceæ; the Cactaceæ of America are represented by certain succulent forms of Mesembryanthemaceæ and Euphorbiaceæ in Africa; and by some Crassulaceæ in Europe. Trientalis europæa has a representative form in America, T. americana; Cornus suecica occurs in Europe, C. canadensis in Canada. Empetrum nigrum, in Arctic regions, has E. rubrum to take its place in the antarctic; Pinguicula lusitanica, in the northern hemisphere, has P. antarctica closely resembling it in the southern; Hydnora africana and H. triceps in South Africa are represented in South America by H. americana.

'The mode in which the globe has been clothed with vegetation has given rise to much discussion. We know from the Sacred Record, that on the third day of the Creation the earth brought forth grass, and herb yielding seed after his kind, and the tree yielding fruit after his kind; but whether the whole earth was at once clothed with vegetation, or certain great centres were formed, whence plants were gradually to spread, we have no means of knowing. The endemic limitation of certain orders, genera, and species, would certainly lead to the opinion, that, in many instances, there have been definite centres, whence the plants have spread only to a certain extent. But the general distribution of other tribes of plants, and

the occurrence of identical species in distant parts of the
world, would favour the view that countries with similar
climates had originally many species of plants in common.
In the case of grasses, we would naturally suppose that
they must have been produced in their social state,
forming pasture for the nourishment of animals; and such
we might conjecture to be the case with social plants in
general.

' Edward Forbes advocates strongly the view of specific
centres, and endeavours to account for the isolation of certain species or assemblages of plants from their centres, by
supposing that these outposts were formerly connected, and
have been separated, by geological changes, accompanied
with the elevation and depression of land. Schouw opposes
this. He thinks that the existence of the same species in
far distant countries is not to be accounted for on the supposition of a single centre for each species. The usual
means of transport, and even the changes which have
taken place by volcanic and other causes are inadequate,
he thinks, to explain why many species are common to
the Alps and the Pyrenees on the one hand, and to the
Scandinavian and Scotch mountains on the other, without
being found on the intermediate plains and hills; why
the flora of Iceland is nearly identical with that of the
Scandinavian mountains; why Europe and North America,
especially the northern parts, have various plants in
common, which have not been communicated by human
aids. Still greater objections to this mode of explanation,
he thinks, are founded on the fact that there are plants at
the Straits of Magalhaens, and in the Falkland and other
antarctic islands, which belong to the flora of the Arctic
pole; and that several European plants appear in New
Holland, Van Diemen's Land, and New Zealand, and
which are not found in intermediate countries. Schouw,
therefore, supposes that there were originally not one, but
many primary individuals of a species.'

A writer in the *Scotsman* has called attention to the

inadequacy of the views of Hutton and Count Saporta to account for past changes of temperature. He states that the theory which now meets with most acceptance is that which attributes climatic changes to astronomical causes, and especially to the eccentricity of the earth's orbit—a theory associated with the name of Dr Croll, of the Geological Survey of Scotland. At the same time, in regard to the facts in which we are here concerned, he says:—

'Besides the tale of intolerable cold brought home by Arctic expeditions, which, in the concrete form of thick-ribbed ice and perpetual snow, has hitherto effectually barred their approach to the Pole, they seldom fail to secure satisfactory evidence of the former existence of more genial conditions in circumpolar lands. Pine trees have been found prostrate on the site of their growth in greatly higher latitudes than those in which they could now exist, and these, from their still unfossilised condition, give evidence of having lived and died within comparatively recent times. That much warmer conditions prevailed at a still earlier time is evidenced by the fossil plants found in some of the highest lands yet reached. These include many evergreen shrubs, oaks, maples, beeches, poplars, and walnuts; while two species of vines have been found fossil in Greenland; *sequoias*, allied to the mammoth trees of the Yosemite region of California in Spitzbergen; with water lilies and the swamp-cypress of the Southern United States in Grinnell Land, within eight degrees of the Pole. Judged by its plant remains, Greenland would appear to have possessed in miocene, or, as many geologists are now inclined to believe, in eocene times, a climate as warm as that of New York or St. Louis, and a vegetation richer than that of Southern Europe at the present time, while the Pole itself, or at least its near neighbourhood, would probably have compared favourably in climate and vegetation with Scotland of to-day. Geologists are agreed in regarding the presence of such a flora as conclusive evidence of the former existence of a warm polar climate; hitherto, however, they have been, and to some extent still are, divided on the question of accounting for it. Some have attributed it to the passage of the solar system through a warmer region of stellar space, others to alteration in the position of the poles. . . . . Abundant evidence of the occurrence in Arctic and Subarctic regions of a series of warm periods extending as far back as the Silurian times, is found in the fossils of the various formations represented in their strata; and the remarkably complete succession of fossil floras there met with, and their marked resemblance to those of lower latitudes, forms one of Saporta's arguments in favour of his view that the circumpolar area has been the birthplace of plants and the centre of their dispersal or migration, a theory which, in its main features, has received remarkable corroboration from the recent investigations of Dawson, Dyer, and Gardner. The rich vegetation of circumpolar lands in eocene times, migrated southward as the climate gradually grew colder, giving place to the modern Arctic flora, which in turn crept slowly southward as the cold of the glacial epoch became gradually more intense, until at length a truly Arctic flora abounded in Central Europe. As the climate slowly ameliorated, the Arctic plant, in order to find suitable conditions, migrated northwards unless where the presence of mountains enabled them to obtain the necessary cold by climbing upwards instead of polewards, and the present alpine flora of the Pyrenees, the Alps, Britain, and Scandinavia, chiefly resembling as it does the vegetation of the Arctic regions, is, as Professor Geikie recently expressed it in his lecture on *Geographical Evolution*, "a living record of the ice age."'

CHAPTER III.

FAUNA.

In Northern Russia, with the exception of insects, the number of species of animals is limited, and they are only such as are found generally in almost all the countries of the same latitude.

SECTION I.—QUADRUPEDS.

Game is plentiful in Russia. There are plenty of hares, *Zayatz:* the white hare, which frequents the woods and moors, weighing from seven to ten pounds, and the red hare of the plains and cultivated lands, weighing from ten to fifteen pounds; and rabbits, *Krolik,* are abundant.

By the end of September shooting with dogs is considered over for the season; the capercailzie and black game have retired to the thickest woods; the willow grouse are packed, and defy the most wary dog; and the snipe and woodcock have all left for warmer climes.

Battue shooting now commences, and though a large head of game is seldom bagged, the sportsman finds variety in the game driven before him to be shot, and a wildness in the vast woods and moorland which possess a charm peculiarly their own!

There are bears, *Medred,* to be found in almost all the forests of Russia.

Bear hunting in Russia is generally conducted thus:—
'As soon as the first snow falls the peasants go out in search of bear tracks. When they come upon traces they follow the track, until they know by the numerous turns and twists which Bruin has made that he is thinking of choosing some snug corner for his winter quarters; they

then proceed with greater caution, and when they consider that the bear is not very far off they leave the track and make a circle, returning to their starting place. Should they, when making this ring, again cross the track of the animal, they know that he has gone beyond the space they are enclosing, and therefore, instead of returning to the starting point, they follow the fresh track and proceed as before. If they do not again cross a track they know that the bear must be within the circle; they advance a little way within the circle, and make another ring; and thus they proceed, gradually limiting the circle, until they have enclosed the bear within a comparatively small circumference. They next inform the sportsmen of what they have done, beaters are then collected, the number varying according to the extent of the circle; they are placed in a semi-circle, while the sportsmen stand in a line at distances from fifty to eighty yards from one another, according to the number of guns, and the nature of the ground. The bear, roused from his slumbers by the shouts and cries of the peasants, generally comes within shot of one or another of the guns, but it seldom happens that a single shot suffices to kill *Misha* (Michael), as the Russians call him. When wounded, the bear, especially if it be a mother with cubs, is dangerous to encounter; the sportsman is generally, however, provided with two guns, and with a spear to be used *en dernier ressort*. In other cases a peasant having discovered later in the season where the bear has made his den, gives information of this to the sportsman, who, along with his informant, and it may be a friend, betakes himself to the spot, generally taking with him three or four rough dogs, which answer the double purpose of rousing the bear from his lair, and of distracting his attention from the sportsman. Some bear hunters make a regular campaign for several weeks together, camping out at night in the forest, and pursuing, it may be for days in succession, a bear likely to escape them.

'The best season of the year for bear-hunting is January

and February, at which time the snow is in a favourable condition for snow-shoes being used, without which the hunter might sink deep in the snow, and would be powerless in following up a bear.

'The snow shoes are about seven feet long and six inches broad, slightly curved at the point, with a foot piece in the middle, to which are attached thongs or straps for securing the snow-shoe to the foot. Some of them are covered underneath with the skin of the reindeer, which is of great assistance to the hunter in ascending hills. In the absence of this undercovering of skin, the hunter provides himself with a pole about eight feet in length, with a curved point of horn or bone, with which he guides himself in descending, or prevents his feet from slipping backwards in ascending, any rising ground.'

In Nova Zembla the Polar bear is met with, and in Northern Russia the glutton or wolverine, which is nearly allied to the *Ursus Arctos*. It is known as the *Taxus gulo*, and as *Gulo Arcticos*. It is apparently the *Ursus luscus* of Linnæus.

The glutton, or wolverine, owes its popular name to its extreme voracity; but it is not less characterised by its strength, fierceness, and cunning. It does not hesitate to dispute their prey with the wolf and the bear, and it baffles frequently the stratagem of the hunter. It is slow and somewhat unwieldy in its movements, but it is determined and persevering, and will proceed at a steady pace for miles in search of prey, stealing unawares upon hares, marmots, and birds; and surprising even the larger quadrupeds, such as the elk and the reindeer, when asleep.

All of the other families of the carnivora, the *Felidae* or cat tribe, the *Canidae* or dog tribe, and the *Mustilidae* or weasel tribe, have their representatives here.

One representative of the *Mustilidae* is the ermine, and another is the sable. The mention of these suggests also the fact that there are squirrels of different species; and beavers have been found in some parts of Lapland—some of these have been found white in colour, but the instances are rare.

A representative of the *Felidae* is the lynx. The lynx is occasionally shot in the vicinity of St. Petersburg. The species most generally found is the *felis vergata* of Nilssen. He is a very wary animal, and even when *ringed* is very difficult to drive from his lurking place.

The *Canidae* are represented by the wolf and the fox.

In Lapland the foxes are extremely numerous, some of them are white with black ears and feet, some red, or red with a black cross, some black, or black with long hairs on their back of a silver colour: the skins of these are highly valued in the north of Europe.

The Arctic fox (*lagopus*, or *isatis canislagopus*), is a dog-like animal, considered native to Spitzbergen and Greenland, but it extends over all the Arctic regions of America and Asia, and it has been found in Finland and Northern Russia.

The wolf (*Volk*) is both abundant and widely diffused. Wolves are shot by hunting them with dogs or by an ordinary *battue*, such as has been described in connection with the hunting of the bear, and sometimes by riding them down, but this requires a peculiar condition of the snow and appropriate ground.

In hunting the wolf it is not uncommon for the sportsman to take with him in his sledge a young squeaking pig in a bag; and a bag of hay of like size is attached to the sledge to be trailed behind. From time to time the pig is pinched—its squeaking attracts wolves—they, seeing the bag of hay trailed behind, supposing the sound to proceed from it, come out to reconnoitre, and thus present themselves within reach of the sportsman's rifle.

The wolves also have their little stratagems. A member of my family, resident for some time in the locality, told me that he had himself seen indications of the following trick, which is not uncommon with them. A herd of wolves, coming to a village, placing themselves in ambuscade, send a she-wolf of the herd yelping through the village, running at full speed, and returning again with like speed and yelping still. The dogs of the village are roused, give chase, and when beyond the ambuscade are attacked by the herd, killed, and devoured. In the case referred to, my son-in-law was informed by a peasant that he had heard the yelping, and on going to the village he saw the footprints of the dogs and the wolves. The ambuscade appeared from the footsteps to have been at the back of the first house in the village; thence they could trace by blood drops on the snow the route by which the wolves had carried off their prey; and on reaching their rendezvous they found shreds of the skins of the dogs lying about like hides in a tan-yard.

In Western Siberia are reared large herds of horses. To these the wolves are very hurtful. In travelling there in the summer there may be seen large herds of mares with their foals grazing, accompanied by a stallion, who acts as *paterfamilias;* should a wolf make his appearance he drives the mares into a circle, with the foals in the centre, and the mares looking towards them; and while he rages and defies the wolf, they are ready to receive him as a square of infantry receives a charge of cavalry, should he venture within reach of their hoofs.

A statistical report lately addressed to the Minister of the Interior estimates the damage done by wolves in 45 European Governments of Russia during the year 1873 at $7\frac{1}{2}$ millions of roubles. The Government of Samara was set down as the greatest sufferer to the extent of 650,000 roubles; next came Vologda at 560,000 roubles, and so on. The Polish and Baltic provinces and Archangel came off best. But competent judges consider this estimate of wolfish mischief as much too low. It is calculated on the basis of a low average value for all Russia, as if the price of an ox or a sheep was about the same everywhere throughout the Empire. It also sets the absolute amount of mischief at far too low a figure. Probably 15 millions of roubles, or £2,500,000, would more nearly represent the value of the domestic animals destroyed annually by wolves in European Russia. To this should be added the value of the wild animals destroyed by them. The reindeer alone killed in Siberia would represent a high figure. Then there is the loss of human life, which can never be accurately known. In 1875 the police reported 161 persons killed by wolves.

In severe winters wolves have been seen in villages within twelve miles of St. Petersburg, and once or twice I have heard of wolves having penetrated even the capital itself. I also heard of several head of elk being destroyed near Payala, a village near St. Petersburg. The village is twelve miles distant from St. Petersburg; it is on the Finnish railway, and between it and the Gulf of Finland is the forest and hunting lodges of Lachta. In this forest bears, wolves, and elks, are found.

Elk shooting is conducted much in the same way as the ordinary *battue* for bear; but the peasants will sometimes follow them for days for the chance of getting a shot.

My son-in-law gave me the following account of an elk hunt near Ejora, on the Neva, in which he took part:—
' One evening in early winter information was brought by a peasant that a herd or family of elks had been tracked to a small wood some miles distant on the opposite side of the river. A party of six were soon formed to go armed in quest. It was eleven o'clock at night; they crossed the river; traversed some distance, bivouacked for the night; and at early dawn were again astir. Reaching the wood, they found it skirted on the one side by a marsh, frozen over with ice of no great thickness, and measures to drive the elks thither were speedily resolved on. They were accompanied by six peasants, and the twelve men were soon stationed in a semi-circle, with the frozen marsh for a base. By previous concert they gradually contracted

the semi-circle, and at a given signal, raised a shout which startled the elks in their layer. These attempted to break the cordon, but startled by the firing of rifles they turned and made for the marsh; seven were young, and crossed it in safety; two large elks, a male and a female, making the attempt, broke the ice, and floundring, found themselves unable to make their escape. The sportsmen came up to within ten feet, fired, killed them, and went home, when they sent a cart or sledge for the carcases. The elk is known in Russia as *olene*.

To the same family belongs the reindeer (*cervus turandus*) so extensively domesticated in Lapland as to be intimately associated with our every conception of the Laplanders, but found in a wild state both in Northern Russia and in Finland, to the north of 65° 30′, and on the northern slopes of the Maanselka.

In the Eastern Hemisphere the isothermal line of 0° descends towards the 55th parallel of latitude, which is lower than it does in America. But there are some important towns situated to the north of this latitude—Tobolsk, lat. 58° 11′; Jokutsk, lat. 58° 16′; and Yakutsk, lat. 62°. In Europe the only Arctic lands properly so called, and distinguished by an Arctic flora, are Russian Lapland and the deeply-indented coast of Northern Russia, and the former is what may be considered the habitat of the reindeer. But during several winters they have been seen in St. Petersburg, brought thither as curiosities, and attracting attention as they were driven along the Neva by the Lappish owners in their national sledges.

In the regions of the reindeer, in Lapland and in Siberia, as in Labrador and the northern coasts of America, the lemming also is met with.

Section II.—Birds.

Of birds we find in Northern Russia those which are common to different lands in such latitudes, together with some which are peculiar to the country, but these are confined to the southern forest districts and arable regions of the country.

On the northern coast are seen the gossander or dun diver; the smew or white nun, and other species of messenger—*merginas*—a sub-family of the *palmipeds;* the dove-kie, or black gullimot; the eider duck, and tho whistling swan.

The *Cygnus musicus*, or whistling swan, is famous for its migrations. It measures five feet from the tip of the bill to the end of the tail, and eight feet from tip to tip of its extended wings; the plumage is snow-white, with a slight tinge of orange or yellow on the head. Some of these swans winter in Iceland; and it is said that in the long Arctic night their song, as they pass in flocks, falls on the ear of the listener like the notes of a violin. It is an old story that the dying swan sings its own dirge. Tennyson sings in reference to this:—

> ' With an inner voice the river ran,
> Adown it floated a dying swan,
> And loudly did lament—
> The wild swan's death-hymn took the soul
> Of that waste place with joy
> Hidden in sorrow : at first to the ear
> The warble was low, and full, and clear ;
> But anon her jubilant voice,
> With a music strange and manifold,
> Flowed forth a carol free and bold—
> And the creeping mosses and clambering weeds,
> And the willow branches hoar and dank,
> And the wavy swell of the soughing reeds,
> And the wave-worn leaves of the echoing bank,
> And the silvery marish-flowers that throng
> The desolate creeks and ports among,
> Were flooded over with eddying song.'

This is told as what is known of the common domesticated swan, which is also found in a wild state, *Cygnus*,—but it is alleged that the wild swan's voice, even in its death hour, has no such musical sweetness as is thus extolled. It is said to be always harsh and dissonant, and to produce a painful impression when it breaks on the silence of Arctic stillness.

'The lakes of Iceland and its streams,' says the author of *The Arctic World*, 'abound with these beautiful birds. They are very numerous on the Myvatu or Great Lake, where are also seen the wild duck, the scoter, the common gossander, the red-breasted merganser, the scaup-duck, and other anserines. They are found also upon the salt and brackish waters along the coast. It is chiefly at the pairing season, or at the approach of winter, that it assembles in multitudes; and as the winter advances it mounts high in the air, and directs its course in search of milder climes. It is in its flight that the *Cygnus musicus*, apparently by the flapping of the air with its wings, occasions the violin-like music to which reference has been made.

'The female builds her nest of withered leaves and stalks of reeds and rushes in lonely and sequestered places. She usually lays six or seven thick-shelled eggs, which are hatched in about six weeks, when both parents assiduously guard and feed the cygnets.

'The wild swan is shot or caught for its feathers, which are highly prized for ornamental purposes; next to the skin lies a coat of thick fine down of the purest white swan down.'

Flocks of wild swans, wild geese, and wild ducks, find their way to the southern limit of the forest region of Northern Russia and beyond it; and there, in the month of August, are seen large flights of snipes passing, it is supposed, in migration from the district around Archangel to the south. If the weather be fine, each flight seems to rest only one night; when it is otherwise they remain for some days, frequenting marshy ground and streamlets, and many fall by the guns of the peasants and sportsmen.

Of the *Tetraonidae*, or grouse tribe, which seem chiefly to inhabit cold countries, there are numerous species, chief amongst which, as being most abundant and most delicious, is the *riabchik*, or so called Russian partridge, but sometimes called grouse. The wood-cock is called *Teterew*, and the wood-hen *Teterka*. The white grouse, ptarmigan, *Lagopus mutus*, is known as *Kuropatka*; the capercailzie is known as *Glukar*; the hazel grouse, *Tetrao bonasia* as *Pyg*; the quail as *Perepelka*; the partridge as *Metsänka*. The snipe is called *Doupel*; the duck, *Torsa*; wild duck, *Dikaia utka*; wild grouse, *Diki gus*; wild swan, *Diki lebea*.

I have no sympathy with prince, peer, or peasant, who finds sport in wounding, maiming, and killing any animal—fish, fowl or quadruped; nor have I any feeling of great respect for the skill and cunning which they put in exercise to enable them to deceive and disarm in order to destroy; neither do I consider that the subsequent utilisation as food of the carcases of the animals they have killed in sport affects greatly the character of their deed: it was for sport, not for food, that the deed was done. With the barbarous or semi-civilised hunter it is otherwise. With this passing remark, to prevent misapprehension of my views while supplying the preceding details, I proceed.

SECTION III.—INSECTS INJURIOUS TO FOREST TREES.

By Forst-Meister Alexander Günther, of the Forest Circuit of Petrazavodsk, I was supplied with the following observations in regard to the ravages of insects in the Government of Olonetz.

'All the land lying to the west of Lake Onega, from the river Svir in the south, the entire western coast of the Onega, and from it to the town of Povonetz, and thence to the Lake Vig and the river of the same name, and on to the White Sea, with the large peninsula beyond, is a district characterised by its fauna and flora. Of the propriety of this statement I became fully satisfied after I visited the Onega lake; and a brief glance at a map may suffice to make this intelligible to any one who has himself visited the locality.

'In the west we find the land for the most part cut up by long lakes, and the ground is chiefly rocky; in the east we find fewer lakes, the land is more alluvial, and there are fewer rivers. As a consequence we find the population manifesting some difference in character. In the west there is little of agriculture, but more of hunting, fishing, felling of trees, and floating of timber; and the people go out as carpenters, builders, glaziers, and stone hewers. The eastern coast of the lake is mostly devoted to agriculture, in the Fudoschen circle it is especially that of flax; in the circle of Kargopolsk it is that of rye and other grain. The eastern part raises so much corn, that, excepting in the Wytegorsk circle, there is corn to be sold; but in the western part almost half of what is consumed is purchased.

'In like manner, if we look to the flora, we see also a few differences; we find, as might have been expected, plants common in the west, which are rare in the east; and also the reverse, plants common in the east which are rare in the west. In the east we find as characteristic of the flora, *Larix Siberica, Atragena alpina, Delphinum elatum, Silene tartarica, Pyrethrum cormybosum, Crepis sibirica*, which are all awanting in the west; while in the east there naturally are missed all the plants which grow upon rocks in the west.

'My list of plants was necessarily defective, as I followed only a special object in preparing for it, and my undertaking was limited to the environs of Lake Onega and to the outlying places around.

'My list of *Lepidoptera* is not yet complete, but I may mention here a few of the more injurious of these. Of *Lepidoptera* which have been very hurtful to agricultural produce since 1870, may be mentioned *Agrotis exclamationis* and *segitum*, and, in passing, *Hydrocœia nictitans, Hadena oculea* and *H. basilinea*. These insects do enormous injury; and more especially does the *Characas graminis* do great damage to the meadows, while the caterpillars devour the roots of the grasses on extensive stretches of land.

'Of the injurious forest insects I cannot say much. Though we have a goodly share of these, we do not see much serious injury done by them. As is known, the injurious forest insects are divided into two sections:—1. Insects which reduce the trees to a sickly condition; to this class belong the *Chrysomelidae, Melolonshidae,* many *Curculionidae,* also *Polydrosus nurcans, Strophosonins coryli, Bruchus, Tortrix trobilana, Hylesinus piniperda, Retinia turionana, Resinella buoliana,* and the caterpillars of *Lophyrus.* 2. Then come the insects which entirely kill the trees, or make the wood altogether useless for technical purposes. In this respect the *Bostrychus lineatus* has done great damage of late years. In December 1879 the gales did great destruction by breaking over and uprooting trees. Many thousands and thousands upon thousands of trees were overturned, and the mountain ridges suffered especially: in some places there were whole stretches of forest levelled with the ground. In 1880 the Crown ordered the sale of the fallen trees, but at the prices of standing trees, and there were of course few bought. In consequence of this, the prices were lowered in 1881, and many were purchased and dressed. But still, in the inspection made in the spring of 1882, it was seen that Nature had not failed to do her work in the matter. *Bostrychus lineatus* had created great devastation. It wrought its way into the wood to the depth of two verstchoks—three and a half inches—and from thick trees, eight verstchoks—fourteen inches—in diameter, only boards five and six verstchoks—nine inches and ten and a half inches—in breadth could be obtained. I speak here only of the *Pinus sylvestris.*

'Moreover, in the summer of 1882, I made the observation that the said *Bostrychus lineata* readily attacked sick overturned trees of *Betula* and of *Alnus incana.* Their course in penetrating the *Betula* was irregular; but in the *Pinus sylvestris* it was as follows:'—[The description was accompanied by a diagram representing the cambium to the depth of from seven to eight verstchoks; from the outer circumference proceeded several branching blue lines; and in the branches of these, at varying distances, were black circular and oval spots. These represented a horizontal slice of the trunk of a *Pinus sylvestris;* and the blue ramified lines represented the horizontal course made by the insect, the black spots indicated places at which it departed from this to take a course more or less vertical]. 'According to my observation,' M. Guenther went on to say, "the insect never penetrated beyond the cambium. Besides this, old trees were throughout their whole thickness entirely destroyed by similar oval vertical openings, produced apparently by the *Tetropium luridum.*'

Section A.—Coleoptera.

The following is a list of Coleoptera, according to *Catalogus Coleopterorum Europae et Caucasi*, auctoribus L. V. Heyden, E. Reitler, et J. Weise, collected in the Government of Olonetz by Forst-Meister A. Guenther, Petrazavodsk:—

'*Catalogus Coleopterorum Gubern. Olonetzeus.*

Cicindela campestris, L.   C. hybrida, L.   C. sylvatica, L.
Cychrus caraboides, L.
Melancarabus glabratus, Pk.
Carabus nitens, L.   C. clathratus, L.   C. granulatus, L.
C. Menetriesi, Fisch.
Nebria cursor, Müll.   N. Gyllenhali, Schh.
Leistus ferrugineus, L.   L. rufescens, Fbr.
Notiopholus aquaticus, L.   N. palustris, Dft.   N. biguthatus, Fbr.
Blethisa multipunctato, L.
Elaphrus uliginosus, Fbr.   E. cupreus, Dft.   E. riparius, L.
Tachypus pallipes, Dft.   T. flavipes, L.
Bembidion velox, L.   B. punctulatum, Drap.   B. bi-

punctatum, L.   B. obliquum, Strm.   B. lampros, Hbst.
B. lampros, var. 14 striatum, Tunis.   B. Grapei, Gyll.   B. tenellum, Er.   B. Giloipes, Strm.   B. Schüppeli, Dej.   B. Doris, Pauz.   B. 4 maculatum, L.   B. saxatile. Gyll.   B. femoratum, Strm.   B. rupestre, L.   B. ustulatum, L.   B. Mannerheimii, Saklb.   B. guttula, F.

Tachys nanus, Gyll.

Trechus rubeus, F.   T. 4 striatus, Schrk.   T. secalis, Payr.

Patrobus excavatus, Payr.   P. var. assimilis, Chaud.   P. var. clavipes, Thms.   P. septentrionalis, Dej.

Broseus cephalotes, L.

Clivina fossor, L.   C. collaris, Hbrt.

Dyschirius, globosus, Hbrt.   D. thoracicus, Rossi.   D. aeneus, Dej.

Lorocera pilicornis, F.

Panagaeus cruix major, L.

Oodes helopiodes, F.

Chlaenius tibialis, Dej.   C. nigricornis, L.

Badister bipustulatus, Fbr.

Anisodactylus binotatus, Fbr.

Ophonus puncticollis, Payr.   O. brevicollis, Serv.

Pseudophonus pubescens, Mull.

Harpalus aeneus, F.   H. aeneus, v. confusus, Dej.   H. discoideus, Fbr.   H. aubripes, Dft.   H. latus, L.   H. latus, v. erythrocephalus, F.   H. luteicornis, Dft.   H. tardus, Pauz.   H. tardus, v. angustior, F. Sahlb.

Bradycellus collaris, Payr.

Stenolophus dorsalis, F.

Amara plebeja, Gyll.   A. littorea, Thms.   A. similata, Gyll.   A. ovata, Fbr.   A. nitida, Ptrm.   A. communis, Pauz.   A. nigricornis, Thms.   A. lunicollis, Schdt.   A. curta, Dej.   A. aenea, Dej.   A. famelica, Zimm.   A. eurynota, Pauz.   A. familiaris, Dft.   A. lucida, Dft.   A. tibialis, Payr.   A. ingenua, Dft.   A. erratica, Dft.   A. interstitialis, Dej.   A. livida, F.   A. praeternussa, Saklb.   A. aulica, Pauz.   A. consularis, Dft.   A. fulva, Deg.   A. apricaria, Payr.

Plerostichus oblongopunctatus, F.   P. vitreus, Dej.   P.

angustatus, Dft. P. niger, Schall. P. vulgaris, L. P. nigritus, Fbr. P. gracilis, Dej. P. minor, Gyll. P. strenuus, Pauz. P. diligens, Strm.

Poccilus dimidiatus, Olio. P. Koyi, Germ. P. lepidus, Leske. P. cupreus, L. P. coerulescens, L.

Lagarus vernalis, Pauz.

Laemostenus inaequalis, Pauz.

Calathus fuscipes, Goez. C. erratus, Sahlb. C. ambiguus, Payr. C. melanocephalus, L. C. micropterus, Dft.

Synuchus nivalis, Pauz.

Platynus Mannerheimi, Dej.

Agonum, 6 punctatum, L. A. Mulleri, Hbrt. A. viduum, Pauz. A. viduum, v. moestum, Dft. A. dolens, Sahlb. A. 4 punctatum, Deg.

Europhilus piceus, L. E. gracilus, Gyll. E. fuliginosus, Pauz. E. fuliginosus, v. puellus, Dej.

Lebia chlorocephala, Hoffm   L. cruix minor, L.

Cymindis macularis, Dej.   C, vaporariorum, L.

Dromius sigma, Rossi.

Metabletus truncatellus, L.

Haliplus variegatus, Strm.   H. ruficollis, Deg.   H. fluviatilis, Aub.

Noterus crassicornis, Müll.

Laccophilus obscurus, Pauz.

Bidessus parvulus, Mull.

Hyphydrus ferrugineus, L.

Coelambus inaequalis, F.   C. versicolor, Schall.   C. 5 lineatus, Zett.   C. impressopunctatus, Schall.

Deronectes brevis, Strm.

Hydroporus lineatus, F.   H. minimus, Scop.   H. melanarius, Strm.   H. nigrita, Fbr.   H. obscurus, Strm.   H. pubescens, Gyll.   H. glabriusculus, Aub.   H. notatus, Strm.   H. neglectus, Schaum.   H. scalesiamus, Stft.   H. angustatus, Strm.   H. umbrosus, Gyll.   H. striola, Gyll., v. vittula, Er.   H. palustris, L.   H. erythrocephalus, L.

Agabus affinis, Payr.   A. unguicularis, Thms.   A. congener, Payr.   A. clypealis, Thms.   A. Mimmii, F. Sahlb.

COLEOPTERA.

A. confinis, Gyll.   A. Sturnii, Gyll.   A. chaleonotus, Pauz.
A. Erichsoni, Harold.   A. melanarius, Aub.
Platambus maculatus, L.
Flybius ater, Deg.   F. obscurus, Marsch.   F. subaeneus, Er.   F. crassus, Thoms.   F. gutliger, Gyll.   F. aenesceus, Thms.   F. angustior, Gyll.   F. fenestratus, F.
Rhantus Grapei, Gyll.   R. notaticollis, Aub.   R. suturalis, Laiod.   R. exoletus, Forst.   R. bistriatus, Bergsk.
Colymbetes Paykulli, Er.   C. striatus, L.
Dyticus marginalis, L.   D. circumcinctus, Ahr.   D. latissimus, L.
Acilius sulcatus, L.   A. fasiatus, Dej.
Gyrinus minutus, F.   G. natator, L.   G. dorsalis, Gyll., v. maximus, Gyll.   G. dorsalis, v. opacus, Saklb.
Hydrochus brevis, Hbst.
Hydraena riparia, Kugel.
Helophorus aquaticus, L.   H. frigidus, Graels.   H. strigifrous, Thms.   H. aeneipennîs, Thms.   H. granularis, L.   H. brevipalpis, Bedel.   H. nanus, Strm.
Hydrobius fuscipes, L.
Creniphilus globulus, Payr.
Philydrus melanocephalus, Ol.   P. testaceus, F.   P. 4 punctatus, Hbst.   P. frontalis, Er.   P. minutus, F.
Laccobius minutus, L.
Chaetarthria semimulum, Pr.
Limnobius truncatellus, Fbr.
Sphaeridium scarabaeoides, L.   S. bipustulatum, F.
Coelostoma orbiculare, F.
Cercyon ustulathm, Preysth.   C. impressus, Strm.   C. melanocephalus, F.   C. haemorrhoidalis, F.   C. lateralis, Marsch.   C. marinus, Thms.   C. bifenestratus, Kust.   C. unipunctatus, L.   C. quisquilius, L.   C. analis, Payr.   C. lugubris, Payr.
Megasternum Colitophagum, Marsch.
Cryptopleurum minutum, F.
Limnichus pygmaeus, Strm.
Dryops prolifericornis, F.   D. auriculata, Pauz.
Limnius Dargelasi, Latr.

Elmis changei, Latr.
Latelmis Volkmari, Pauz.
Georyssus crenulatus, Rossi.
Bolitochara lunulata, Fr.
Leptusa analis, Gyll.
Notothecta confusa, Maerk.
Aleochara fuscipes, Grav. A. brevipennis, Grav. A. morion, Grav. A. mycetophaga, Kr. A. lanuginosa, Grav. A. moereus, Gyll.
Dinarda dentata, Grav.
Lomechusa strumosa, F.
Myrmedonia collaris, Payr. M. cognata, Maerk. M. limbata, Payr. M. laticollis, Maerk.
Astilbus canaliculatus, F.
Falagria suculata, Payk. F. obscura, Grav.
Tachyusa coaretata, Er. T. leucopus, Marsh.
Gnypeta carbonaria, Marsh.
Homalota longula, Heer. H. debilis, Er. H. gemina, Er. H. arctica, Thoms. H. elongatula, Grav. H. melanocera, Thms. H. Gyllenhali, Thms. H. graminicola, Grav. H. silvicola, Fuss. H. euryptera, Steph. H. brinotata, Kr. H. xanthopus, Thms. H. sericans, Grav. H. picipennis, Muuh. H. atramentaria, Gyll. H. longicornis, Gyll. H. excellens, Kr. H. circellaris, Grav. H. exilis, Er. H. talpa, Heer. H. analis, Grav. H. sordida, Marsch. H. pygmaea, Grav. H. aterrima, Grav. H. fungi, Grav. H. laticollis, Steph.
Placusa humilis Er. P. infima, Er.
Thectura aequata, Er. T. angustata, Gyll. T. plana, Gyll.
Hygronoma dimidiata, Er.
Dasyglossa prospera, Er.
Oxypoda opaca, Grav. O. lentula, Er. O. haemorrhoa, Sahlb. O. annularis, Sahlb.
Ocyusa maura, Er.
Gyrophaena laevipennis Kr.
Ollgota pusillima Grav.
Myllaena dubia, Grav. M. minuta, Grav.

Dinopsis erosa, Steph.
Tachinus 'rufipes, L. T. pallipes, Grav. T. marginatus, Gyll. T. proermus Kr. T. flavipes, F. T. laticollis, Grav. T. marginellus, F. T. collaris, Grav. T. subterraneus, L. T. fimetarius, F.
Tachyporus obtusus, L. T. abdominalis, Gyll. T. pallidus, Scharp. T. chrysomelinus. T. jocusus. Say. T. hypnorum, F. T. transversalis, Grav. T. macropterus, Steph.
Conurus pubescens, Payr. C. pedicularis, Grav.
Bolitobius lanulatus, L. B. trimaculatus, Pr. B. pygmaeus F.
Megacrous cingulatus, Marsch. M. formosus, Grav.
Mycetophorus splendidus, Grav. M. elegans, Maeal.
Quedius mesomelinus, Marsh. Q. xanthopus, Er. Q. laevigatus, Gyll. Q. fuliginosus, Grav. Q. molochinus, Q. obliteratus, Er. Q. attenuatus, Gyll.
Emus maxillosus, L.
Leistotrophus murinus L.
Haphyllnus pubescens, Dej. H. erythropterus, L. H. caesareus, Cedex. H. latebricola, Grav. H. fulvipes, Scop. H. picipennis, F. H. fuscatus, Grav. H. aeneocephalus, Deg.
Actobius cinerasceus, Grav.
Philanthus nitidus. F. P. splendens, F. P. atrata. Grav. P. aeneus, Rossi. P. cephalotes, Grav. P. corrinus, Er. P. sanguinolentus, Grav. P. laminatus, Creutz. P. rotundicollis, Menet. P. fimetarius, Grav. P. nigritulus, Grav. P. splendidalus, Grav. P. vernalis, Grav. P. politus, F. P. varius, Gyll. P. marginatus, Mull. P. crueutatus, Gmel. P. varians, Payr. P. agilis, Grav. P. lepidus, Grav. P. nigrita, Grav. P. micans, Grav. P. tenuis, F.
Othius myrmecophilus, Kiesw. O. lapidicola, Kies.
Baptolinus pilicornis, Pr.
Leptacinus botychrus, Gyll. L. formiscetorum, Malen.
Xantholinus tricolor, F. L. linearis, Olio. L. punctulatus, Pr. L. ochraceus, Gyll. L. lentus, Er.
Lathrobium elongatum, L. L. boreale, Hochh. L. ful-

vipenne, Grav. L. fovalum, Steph. L. quadratum, Payr.
L. atripalpe, Scrib.
   Stilicus rufipes, Germ.
   Sunius neglectus, Maerk.
   Paederus riparius, L.
   Dianous coerulescens, Gyll.
   Stenus bigattatus, L. S. bipunctatus, Er. S. clavicornis, Scop. S. providus, Er. S. lustrator, Er. S. proditor, Er. S. Juno, F. S. fasciculatus, T. Sahlb. S. melanarius, Steph. S. bupthalmus, Grav. S. canaliculatus, Gyll. S. palposus, Zett. S. nitens, Steph. S. fuscipes, Grav. S. argus, Grav S. opticus, Grav. S. scabriculus, F. Sahlb. S. crassus, Steph. S. tarsalis, Ljungh. S. similis, Hbrt. S. cincindeloides, Schall. S. fornicatns, Steph. S. pubescens, Steph. S. binotatus, Jungh. S. palustris, Er.
   Euaesthetus bipunctatus, Ljung. E. laeviusculus, Marsh.
   Oxyporus rufus, L.
   Bledius fracticornis, Payr. B. subterraneus, Er. B. pallipes, Grav.
   Plabysthetus nadifrous, Sahlb. P. arenarius, Fourer.
   Oxytelus rugosus, Grav. O. fulvipes Er. O. scilptus, Grav.
   Hapladeres callatus, Grav.
   Trogophloeus corticinus, Gr. T. foveolatus, Sahlb.
   Anthophagus omallenus, Zett. A. caraboides, L.
   Geodromicus plagiatus v. nigritus, Audl.
   Lestesa longelytrata, Goez.
   Orochares angustata Er.
   Olophrum consimili, Gyll.
   Arpedium quadrum, Grav. A. brachypterum. Grav.
   Homalium pusillum, Grav. H. lapponicum, Zett. H. rufipes, Fourer.
   Anthobium minutum, F. A. longipenne, Er.
   Protinus brachypterus, F. P. mairopterus Gyll.
   Megarthrus depressus, Payr.
   Phlaeochares subtilissima, Marsh,
   Bryadis fossulata, Rohb.
   Rybaxis sanguinea, L.

Bythinus bulbifer, Rohb.
Pselaphus Heisii, Hbrt.   P. dresdensis, Hbrt.
Euplectus signatus, Rehb.   E. ambiguus, Rehb.
Euthia seydmaenoides, Stph.
Scydmoenus collaris, Müll.
Euconnus hirticollis, Fll.
Ptomaphagus morio, F.   P. affinis, Steph.
Colon bidentatum, Sahlb.   C. serripes, Sahlb.   C. serripes, v. puncticollis, Kr.   C. appendiculatum, Sahlb.
Phoxphuga atrata, L.   P. opaca, L.
Thonatopholus thoracicus, L.   T. sinuatus F.   T. rugosus, L.
Necrades littoralis L.
Necrophorus vespillo, L.   N. vestigator, Herschel.   N. investigator, Zett.   N. vespilloides, Herbt.
Hydnobius spinipes, Gyll.
Liodes obesa, Schmdl   L. dubia, Kugel.   L. furoa, Er.   L. ovalis, Schmdt.   L. badia, Sturn.   L. parvula, Sahlb.
Cyrtusa subtestacea, Gyll.
Anisotoma humeralis, Kugs.   A. axillaris, Gyll.   A. glabra, Kugl.
Amphicyllis globus, F.
Agathidium laevigatum, Er.
Plenidium formicetorum, Kr.
Millidium minutissimum, Ljungh.
Trichopteryx sericans, Heer.   T. lata, Mot.   T. thoracica, Waltl.
Orthoperus brumipes, Gyll.
Scaphisoma agaricinum, L.
Phalaerus substriatus, Gyll.
Olibrus aeneus F.   O. bicolor, F.   O. affinis, Strm.
Dacne bipustulata, Thub.
Combocerus glaber, Schall.
Triplax russica, L.   T. aenea, Schall.
Endomychus coccineus, L.
Myrmecoxemis subterraneus, Chir.
Antherophagus pallens, Olio.
Cryptophagus acutangulus, Gyll.   C. scanieus, L.

Atomaria fuscipes, Gyll.
Lathridius lardarius, Deg.
Enicmus minutus, L.
Corticaria pubescens, Gyll.   C. longicollis, Zett.
Melanopthalma gibbosa, Hbst.   M. fuscula, Gyll.
Fritoma picea, F.
Micropeplas tesserula, Curt.
Cerrus bipustulatus, Payr.
Brachypterus urticae, F.
 Erurea aestiva L.   E. variegata, Hbst.   E. obsoleta, F.
E. boreella, Zett.   E. pygmaea, Gyll.   E. pusilla, Fll.
Nitudula, bipustulata, L.
Omosita depressa, L.
Soronia grisca, L.
Meligethes rufipes, Gyll.   M. lumbaris, Strm.   M. brassicae, Scop.   M. subrugosus, Gyll.   M. viduatus, Strm.   M. pedicularius, Gyll.
Pocadius ferrugineus, F.
Cychramus luteus, F.
Tps quadripunctatus, Olio.   F. quadripustulatus, L.
Rhizophagus ferruginus, Payr.   R. dispar, Payr.   R. parvulus, Payr.
Tenebrioides mauritanicus, L.
Ortoma grossum, L.   O ferrugineum, L.   O. oblongum, L.
Orthocerus muticus, L.
Synchitades crenata, F.
Cerylon histeroides, F.
Hyliota plana, L.
Silvanns, unidentatus, Olio.
Monotoma angusticollis, Gyll.
Byturus sambuci, Scop.
Dermester murinus, L.   D. lardarius, L.
Antbrenus museorum, L.   A. fuscus, Ltr.
Cistela pilula, L.   C. fasciatus, F.   C. pustulatus, Fort.
Citela varia, F.
Pedilopherus aenus, F.
Platysoma oblongum, F.
Hister unicolor, L.   H. cadaverinus, Hoffm.   H. succi-

cola, Thms. H. neglectus, Germ. H. carbonarius, Tel. H. ventralis, Mars. H. purpurascens, Hbt. H. bissexstriatus, F.
Saprinus rugifer, Payr. S. aeneus, F.
Systenocerus caraboides, L.
Ceruchus chrysomelinus, Hshw.
Sinodendron cylindricus, L.
Onthophagus michicornis, L.
Asphadius subterraneus, L. A. fossor, L. A, haemorrhoidalis, L. H. fimetarius, L, A. ater, Deg. A. granarius, L. A. patridarius, Hbst. A. sordidus, F. A. rufus, Moll. A. niger, Pauz. A. inquinatus, F. A. conspurcatus, L. A. pusillus, Hbt. A. merdarius, F. A. prodromus, Brahm. A. punctatosuleatus, Hm. A. rufipes, L. A. luridus, Payr. A. depressus, Kugel.
Aegialia sabuleti, Payr.
Geotrupes stercorarius, L. G. spiniger, Marsch. G. sylvaticus, Pauz. G. vernalis, L.
Trox scaber, L.
Servia brunnea, L.
Rhizotrogus solstitialis, L.
Melolontha hippocastani, F.
Phyllopertha horticola, L.
Anomala aenea, Deg.
Cetonia floricola, Hbst. C. aurata, L.
Trichius fasciatus, L.
Chalcophora mariana, L.
Dicerca furcata, Thms.
Buprestis rustica, L. B. haemorrhoidalis, Hbt. B. 8 guthata, L.
Melanophila acuminata, Deg.
Phaenops cyanea, F.
Anthaxia 4 punctata, L.
Chrysobothris chrysostigma, L.
Agrilus viridis, L. A. viridis v. nocious, Ratzb.
Trachys minuta, L.
Avelocera conspersa, Gyll. A. fasciata. L.
Lacon murinus, L.

S

Elater cinnabarinus, Esch. E. elongatulus, F. E. balteatus, L. E, tristis, L. E. nigrinus, Payr.
Cryptohypnus riparius, F. C. 4 pustulatus, F.
Cardiophorus ruficollis, L. C. ebeninus, Germ.
Melanotus cartanipes, Payr. M. rufipes, Hbrt.
Limonius aeruginosus, Olio. L. aeneoniger, Deg.
Athous undulatus, Deg. A. subfuscus, Müll.
Corymbites pectinicornis, L. C. cupreus, F. C. cupreus v. aeruginosus F. C. castaneus, L. C. affinis, Payr. C. quercus, Gyll. C. tesselatus, L. C. impressus, F. C. nigricornis, Pauz. C. melancholicus F. C. aeneus, L. C. cruciatus L. C. costalis, Payr.
Agriotes lineatus, L. A. obscurus, L.
Dolopius marginatus, L.
Sericus brunneus, L.
Adrastus pallens, F.
Denticollis linearis, L.
Dascillus cervinus, L.
Helades minutus, L. H. marginatus, F.
Microcara livida, F.
Cyphon coarctata, Payr. C. nitidulus, Thms. C. padi, L. C, variabilis, Thubg. C. pallidulus, Boh.
Eucinetus haemorrhoidalis, Germ.
Eros aurora, Hbrt.
Platycis minutus, F.
Dictyoptera sanguinea, L.
Lampyris noetilma, L.
Podabrus alpinus, Payr. P. lapponicus, Gyll.
Cantharis violaceus, Payr. C. fusca, L, C. rustica, Foll. C. obscura, L. C. nigricans, Mull. C. rufus, L. C. rufus, v. liturata, Foll. C. fuloscollis, F. C. flavilabris, Foll. C. paludosa, Foll. C. haemorrhoidalis, F.
Rhagonycha pilosa, Payr. R. testacea L. R. femoralis, Brutl. R. pallipes, F. R. elongata, Foll. R. atra, L.
Malthinus bigathulus, Payr. M. punctatus, Fourcr.
Malachius bipustulatus, L. M. aeneus, L.
Attalus cardiacae L.
Dasytes niger, L. D. obscurus, Gyll. D. plumbeus, Mull,

Dolichosoma lineare, Rossi.
Cleroides formicarius, L.
Necrobia violacea, L.
Elateroides dermestoides, L.
Bruchus (Ptinus) fur L.
Anobium denticollis, Pauz. A. pertinax, L. A. domesticum, Fourer. A. paniceum, L.
Ernobius nigrinus, Strm.
Dorcatoma dresdensis, Hbrt.
Dinoderus elongatus, Payr.
Cis boleti, Scop. C. nicans, F. C. Jacquemarti, Mill. C. alni, Gyll.
Rhapalodontus fronticornis, Pauz.
Octotemnus glabriculus, Gyll.
Microzoum tibiale, F.
Bolitophagus reticulatus, L.
Diaperis boleti, L.
Scaphidema metallicum, F.
Gnathocerus cornutus F.
Corticeus fraxini, Kugel.
Upis ceramboides, L.
Tenebrio molitor, L.
Bius thoracicus, F.
Mycetochares flavipes, F.
Lagria hirta, L.
Orchesia fasciata, Payr.
Xlyita laevigata, Hellen.
Scotodes annulatus, Eschh.
Henotrachelus aeneus, Payr.
Notoxus monoceros, L. N. cornutus, F.
Anthicus ater, Pauz. A. luteicornis, Schaum.
Pyrochooa pectinicornis, L.
Mordella perlata, Sulz. M. amleata, L.
Mordellistena humeralis, L.
Anaspes frontalis, L. A. flava, L. A. rufilabris, Gyll.
Meloe violaceus, Marsh.
Calopus serraticornis, L.
Ditylus laevis, F.

Oedemera virescens, L. O. lurida Marsch.
Chrysanthia viridis, Schmdl.
Pytho depressus, L.
Rhinosimus ruficollis, L.
Otiorrhynchus dubius, Strm. O. lepidopterus, F. O. ovatus, L.
Phyllobius maculicornis, Gern. P. glaucus, Scop. P. urticae, Deg. P. argentatus, L. P. piri, L.
Polydrusus tereticollis, Deg. P. fasciatus, Müll. P. flavipes, Deg. P. cervinus, L. P. mollis, Stroew.
Platytarsus echinatus, Bourd.
Strophosomus coryli, F. S. obesus, Marsch.
Brachyderes incanus, L.
Sitona flavesceus, Marsch. S. sulesfrons, Thubg. S. lineatus, L.
Tanemecus palliætus, F.
Hypera pollux, F. H. rumicis, L. H. meles, F. H. polygoni, F. H. elongata, Payr. H. suspiciosa, Hbt. H. murina, F.
Cloonus trisulcatus, Hbt.
Lixus paraplecticus, L.
Hylobius piceus, Deg. H. abietis, L. H. pinastri, Gyll.
Pissodes pini, L. P. notatus, F. P. piniphilus, Hbt.
Grypidius equiseti, F.
Erirrhinus acridulus, L. E. aethiops, F.
Docytomus suratus, Gyll. D. salicinus, Gyll. D. tortrix, L. D. dorsalis, L.
Anoplus plantaris, Naer.
Lyprus cylindricus, Payr.
Cryptorrhynchus lapatti, L.
Magdalis phlegmatica, Hbt. M. duplicata, Germ. M. frontalis, Gyll. M. violacea, L. M. carbonaria, L. M. pruni, L.
Balanobius brassicae, F.
Anthonomus rubi, Hbrt. A. pubescens, Payr. A. incurvus, Pauz. A. rectirostres, L.
Acalyptus carpini, Hbrt.
Elleschus bipunctatus, L.

Miarus campanulae, L.
Nanophyes lythri, F.
Orchestes rusci, Hbrt. O. salicis, L. O. stigmo, Germ.
O. foliorum, Müll. O. decoratus, Germ.
Allodactylus geranii, Payr.
Coeliodes rubicundus, Pr.
Cnemogonus epilobii, Pr.
Cidnorrhinus 4 maculatus, L.
Rhinoncus castor, F. R. pericarpius, L.
Phytobius 4 nodosus, Gyll.
Centorrhynchus marginatus. Payr. C. sulcicollis, Payr.
C. erysimi, F.
Tapinotus sellatus, F.
Baris F. album L.
Ryncolus ater, L.
Apion cerdo, Gerst. A. urticarium, Hbrt. A. Hookeri, Kerb. A. simile, Kerb. A. viciae, Payr. A. apricans, Hbrt. A. flavipes, F. A. vireus, Hbrt. A. Gyllenhali, Kerb. A. minimum, Hbrt. A. Sundevali, Boh. A. frumentarium, L. A. violaceum, Kirb.
Rhynchiter cupreus, L. R. planirostris, F. R. megacephalus, Grm. R. betulae, L.
Rhinomacer alni, Müll. R. populi, L.
Apaderes coryli, L.
Platyrrhinus laterostris, F.
Tropideres niveirostris, F.
Maerocephalus albinus, L.
Mylabris atomarius, L. M. loti, Payr.
Hylastes ater, Payr. H. cunicularius. Er. H. opacus, H. glabratus, Zett. H. palliatus, Gyll.
Myclophilus piniperda, L.
Scolytus Geoffroyi, Goez.
Crypturgus pusillus, Gyll. C. cinereus, Hbrt.
Tomicus, 6 dentatus, Boern. T. typographus, L. T. acuminatus, Gyll. T. laricis, F. T. suturalis, Gyll. T. chalcographus, L. T. bidentatus, Hbrt.
Dryocoetes autographus, Rtzb.
Tripodendron lineatus, Olio.

Spondylus buprestoides, L.
Stenocorus mordax, Dej. S. inquisitor, L.
Oxymirus cursor, L.
Pachyta lamed, L. P. 4 maculatus L.
Brachyta interrogationalis, L.
Gaurotes virginea, L.
Acmaeops pratensis, Lauh.
Leptura sanguinosa, Gyll. L. livida, F. L. maculicornis, Deg. L. rubra, L. L. virens, L. L sanguinolenta, F. L. 6 maculata, L. L. chrysomeloides, Schrk. L. nigripes, Deg. L. 4 faterata, L. L. melanura L.
Molorchus minor, L.
Criocephalus rusticus, L.
Tetropium luridum, L.
Asemum striatum, L.
Callidium aeneum, Deg. C. violacem, Fairn.
Semanotes undatus, L.
Clytus rusticus, L.
Acanthocinus aedilis, L. A. griseus, F.
Pogonochaerus fasciculatus, Deg.
Lamia textor, L.
Monochammus sartor, F. M. sutor, F.
Saperda populnea, L. S. carcharias, L. S. scalaris, L. S. perforata, Pall
Oberea oculata, L.
Donacia crassipes, F. D. dentata, Hoppe. D. versicolorea, Rrhm. D. aquatica, L. D. bicolor, Zschael. D. obscura, Gyll. D. thalassina, Germ. D. breolcornis, Ahr. D impressa, Payr. D. fennica, Payr. D. semicuprea, Pauz. D. simplex, F.
Plateumoris sericea, L. P. discolor, Pauz. P. braccata, Scop.
Zeugophora subspinosa, F.
Lema cyanello, L. L. melanope, L.
Crioceris merdigera, L.
Labidostmis 3 dentata, L.
Clytra 4 punctata, L.
Gynandrophthalma saliciva, Scop.

COLEOPTERA.

Cryptocephalus coryli, L. C. 8 punctatus, Scop. C. 6 punctatus, L. C. distinguendus, Schneig. C. bipunctatus, L, C, sericeus, L. C. hypochoeridis, L. C, nitidus, L. C. pini, L. C. labratus, L. C. exiguus, Schneid. C. Moraei, L.

Pachybrachys hieroglyphicus, Satlb.
Pachnephorus tesselatus, Dft.
Adoxus obscurus, L.
Gastroidea viridala, Dej. G. polygoni, L.
Chrysomela staphylea, L. C. marginata, L. C. analis, L. C. varians, Schall. C. fatuosa, L. C. graminis, L. C. polita, L.
Orina gloriosa, F.
Phytodecta rufipes, Dej. P. viminalis, L. P. 5 punctata, F. P. pallida, L.
Phyllodecta vulgotissima, L. P. vitellinae, L.
Hydrothassa aucta, F. H. marginella, L. H. hannoverana, F.
Prasocuris phellandrii, L.
Phaedon armoraciae L. P. cochleariae, F.
Plagiadera versicoloria, Satlb.
Melasoma aeneum, L. M. collare, L. M. lapponicum, L. M. lapponicum, v. bulgœarense, F. M. tremulae, F. M. populi, L. M. longicolle, Suffr.
Phyllobrotica 4 maculata, L.
Luperus pinicola, Dft. L. flavipes, L.
Lochmaea capreae, L.
Trirhabda viburni, Payr.
Gallerucella nymphaeae, L. G. aquatica, L. G. lineola, E. G. calmariensis, L. G. tenella, L.
Galerma tanaceti, L. G. pomonae, Scop.
Crepidocera femorata, Gyll. C. nitidula, L. C. ullxines, L. C. ferruginea, Scop.
Mantura rustica, L. M. chrysanthemi, Koch.
Chaetoenema concinna, Marsch. C. aridula, Gyll. C. Sahlbergi, Gyll. C. hortensis, Fourer.
Psylloides culcullata, Fll.
Haltica lythri, Aub. H. oleracea, L.

Batophila rubi, Payr.
Phyllotreta undulata, Kutsch.
Longitarsis apricalis, Beca. L. luridus, Scop. L. atricapillus, Dft. L. melanocephalus, Deg.
Dibolia cynaglossi, Koch.
Cassida sanguinosa, Suffr. C. vibex, L. C. chloris, Suffr. C. sanguinolata, Mull. C. nobilis, L. C. nebulosa, L. C. flaveola, Thub.
Hippodamia 13 punctata, L. H. 7 maculata, Deg.
Adonia variegata, Gaez.
Anisostica 19 punctata, L.
Adalia bothnica, Payr. A. bipunctata, L.
Coccinella 7 punctata, L. C. trifasciata, L. C. 5 punctata, L. C. hieroglyphica, L. C. 14 pustulata, L.
Mysia oblongogutteta, L.
Halyzia ocellata L. H. 14 gutteta, L. H. 22 punctata L. H. conglobata.
Coccidula rufa, Hbrt.
Chilocorus similis, Rossi. C. bipustulatus, L.
Exochomus 4 pustulatus, L.
Scymnus suturalis, Thub. S. bipunctatus, L.

SUB-SECTION B.—LEPIDOPTERA.

LEPIDOPTERA of the Government of Olonetz, according to *Catalog der Lepidopteren des Europaeischen Tannengebiets von O. Standings, u. M. Wocke*, collected by Forst-Meister Guenther, Petrazavodsk.

Macrolepidoptera.

Papilionidae,	-	-	-	70	
Sphingidae,	-	-	-	22	
Bombycidea,	-	-	-	63	
Noetuidae,	-	-	-	146	
Geometridae,	-	-	-	162	= 463

LEPIDOPTERA.

Microlepidoptera.

Pyralidae,	- - - -	68
Tortricidae,	- - - -	177
Tineidae,	- - - -	238
Pterophoridae,	- - -	17 = 500

963
Varietates et abberationes, - 49

Rhophalocera.

Papilio machaon, L.
Parnassius mnemosyne, L.
Aporia crategi, L.
Pieris brassicae, L. P. rapae, L. P. napi, L. P. napi, ab. byronoae, O. P. daplodice, L.
Anthocharis cardamines, L.
Leucophasia sinapis, L.
Colias palaeno, L. C. edusa, F.
Rhodocera rhamni, F.
Thecla betula, L. T. pruni, L. T. rubi, L.
Polyomniatus virgaureae, L. P. hippothoë, L. P. phlaeas, L. P. amphidamas, Esp.
Lycaena argyrotoxus, Bgstr. L. argus, L. L. optilete, Knoch. L. astrarche, Bgstr. L. icarus. Rott. L. eumedon, Esp. L. amanda, Schn. L. Donzelii, B. L. argiolus, L. semiargus, Rott.
Apatura ilia, Schiff, ab. clytie, Schiff.
Limenitis populi, L.
Vannessa C. albuni, L. V. urticae, L. V. jo, L. V. antiopa, L. V. atalanta, L. V. cardui, L.
Melitaea maturna, L. M. aurinia, Rott. M. athalia, Rott.
Argynnis aphirape, Hb. A aphiraphe, v. ossianus, Hbrt. A. aphirape, v. Isabella, Tgstr. A. selene, Schiff. A. euphrosyne, L. A. pales, Schiff. A. amathusia, Esp. A. frigga, Thnb. A. ino, Esp. A. lathonia, L. A. aglaja, L. A. niobe, L. A. adippe, L. A. paphia, L.

Erebia ligea, L.   E. embla, Thnb.   E. euryale, Esp., v. euryloides, Tgstr.
Oeneis jutta, Hb.
Pararge maera, L.   P. hiera, F.   P. aegeria, L.
Epinephele lycaon, Rott.   E. hyperantus, L.
Coenonympha iphis, Schiff.   C. pamphilus, L.   C. tiphon, Rott.
Syrichtus alveus, Hb.   S. malvae, L.
Hesperia lineola, O.   H. sylvanus, Esp.   H. comma, L.
Carterocephalus palaemon, Pall,   C. silvius, Knoch.

Heterocera.

Sphinx pinastri, L.
Deilephila gallii, Rott.   L. elpenor, L.   D. porcellus, L. D. nerii, L.
Smerinthus ocellata, L.   S. populi, L.   S. tremulae, Tr.
Macraglossa stellatarum, L.   M. bombyliformis, O.   M. fucifornis, L.
Trochilium apiforme, Cl.
Sciapteron tabaniforme, Rott.
Sesia spheciformis, Gerning.   S. tipuliformis, Cl.   S. culiciformis, L.   S. formicaeformis, Esp.
Bembecia hylaeiformis, Lasp.
Ino pruni, Schiff.   T. statices, L.
Zygaena scabiosae, Schev.   Z. lonicerae, Esp.
Sarrothripa undulana, Hb., ab. dilutana, Hb.
Earias clorana, L.
Hylophila prasinana, L.
Nola albula Hb. var. karelica, Tugst,   N. centonalis, Hb.
Nudaria senex, Hb.
Setina irrorella, Cl.   S. irrorella, var. Andereggi, H. S.   S. mesomella, L.
Lithosia complana, L.   L. lutarella L.
Encydia cribune, L.
Nemeophila russula, L.   N. plantaginis, L.   N. plantaginis, ab. hospita, Schiff.   N. plantaginis, ab. matronalis, Frr.

Arctia caja, L.
Spilosoma fuliginosa, L. S. mendica, Cl. S. menthastri, Esp.
Hepialus humuli, L. H. velleda, Hb. H. ganna, Hb. H. hecta, L.
Cossus cossus, L.
Psyche unicolor, Hufu. P. hirsutella, Hb.
Epichnopteryx bombycella, Schiff.
Tumea intermediella, Brd.
Orgyia gonostigma, F. O. antiqua, L.
Dasychira selenitica, Esp. D. fascelina, L.
Leucoma salicis, L.
Bombyx crataegi, L. B. populi, L. B. lanestris L. B. quercus, L. B. rubi, L.
Lasiocampa potatoria, L. L. illicifolia, L. L. pini, L.
Endromis versicolora, L.
Saturnia pavonia, L.
Aglia tau, L.
Drepana falcataria, L. D. curvatula, Bkh. D. lacertinaria, L.
Harpyia furcula, L. H. bifida, Hb. H. vinula, L.
Notodonta dictaeoides, Esp. N. ziczac, L. N. torva, Hb. N. dromedarius, L. N. bicoloria, Schiff. N. bicoloria, v. albida, B.
Lophopteryx Sieversi, Meu. L. camelina, L.
Pterostoma palpina, L.
Gluphisia crenata, Esp.
Phalera bucephala, L
Pygaera timon, Hb. P. curtula, L. P. anachoreta, F. P. pigra, Hufu.
Thyatira batis L.
Cymatophora octogesima, Hb. C. or F. C. duplaris, L.
Asphalia flavicornis, L.
Demas coryli, L.
Aeronycta leporina L. A. megacephala, F. A. alni, L. A. psi, L. A. menyanthidis, View. A. auricoma, F. A. abscondita, Fr. A. rumicis, L.
Diphthera ludifica, L.

Agrotis strigula, Thnb. A. polygona, F. A. subrosea, Stph. A. sobrina, Gn. A. augur, F. A. obscura, Brahm. A. pronulia, L. A. hyperborea, Zett. A. collina, B. A. baja, F. A. sincera, Hs. A. speciosa, Hb, var. arctica, Zett. A. cnigrum, L. A. dahlii, Hb. A. brunnea, F. A. conflua, Fr. A. cuprea, Hb. A. plecta, L. A. fennica, Tausch. A. simulans, Hufu. A. exclamationis, L. A. nigricans, L. A. tritici, L. A. tritici, var. eruta, Hb. A. tritici, var. aquilina, Hb. A. segetum, Schiff. A. corticea, Hb. A. vestigalis, Rott. A. prasina, F. A. occulta, L.

Charoeas graminis, L.

Mamestra serratilinea, Fr. M. advena, F. M. tincta, Brkm. M. contigua, Vill. M. thalassina, Rott. M. dissimilis, Knock, Knoch. M. piri, L. M. brassicae L M. Glauca, Hb. M. deutina, Esp. M. trifolii, Rott. M. reticulata, Vill. M. serena, F.

Dianthoecia proxima, Hb. D. proxima, ab. ochrostigma, Ev. D. nana, Rott. D. compta, F. D. cucubali, Fuessl. D. carpophaga, Bkh.

Polia chi, L.

Luperina Haworthii, Curt.

. Hadena porphyrea, Esp. H. adusta, Esp. H. adusta, var. baltica, Hering. H. rubrirena, Fr. H. furva, Hb. H. lateritia, Hufu. H. basilinea, F. H. rurea, F. H. rurea ab. alopecurus, Esp. H. gemina, Hb. H. gemina ab. remissa, Fr. H. didyma, Esp. H. didyma ab. leucostigma, Esp. H. pabulatricula, Brahm.

Dipterygia scabriuscula, L.

Hyppa rectilinea, Esp.

Chloantha polyodon, Cl.

Euplexia lucipara, L.

Naenia typica, L.

Helotropha leucostigma, Hb.

Hydroecia nictitans, Bkh. H. nictitans ab. erythrostigma, Hu. H. micacaea, Esp.

Tapinostola fulva, Hb., ab. fluxa, Tr.

Leucania pallens, L. L. comnia, L. L. conigera, F.

Mithymna imbecilla, F.
Caradrina morpheus, Hufu.   C. quadripunctata, F.   C. cinerasceus, Tugstr.   C. petraea, Tugstr.   C. alsines, Brahm.   C. taraxaci, Hb.   C. palustris, Hb.
Rusina tenebrosa, Hb.
Amphipyra tragopaginis, L.
Taeniscampa gothica, L.   T. gothica, var. gothicina, H.S.   T. incerta, Hufu.   T. optima, Hb.
Panolis piniperda, Pauz.
Pachnobia rubricosa, F.
Cosmia paleacea, Esp.
Dyschorista suspecta, Hb.
Cleoceris viminalis, F.
Orthosia circellaris, Hufu.   O. helvola, L.
Lanthia flavago, L.   L. fulvago, L.   L. fulvago, ab. flavescens, Esp.
Orrhodia vaccinii, L.   O. vaccinii, ab. polita, F.
Scopelosonia satellitia, L.
Scoliopteryx libatrix, L.
Xylina socia, Rott.   X. ingrica, H.S.   X. lambda, F.
Calocampa vetusta, Hb.   C. solidaginis, Hb.
Asteroscopus nubeculosus, Esp.
Calophasia lunula, Hufu.
Cucullia umbratica, L.   C. gnaphalii, Hb.
Plusia tripartita, Hufu.   P. moueta, F.   P. chrysitis, L.   P. bractea, F.   P. festucae, L.   P. jota, L.   P. macrogamma, Ev.   P. gamma, L.   P. interrogationis, L.   P. microgamma, Hb.
Anarta myrtilli, L.   A. cordigera, Thnb.
Heliaca tenebrata, Sc.
Heliothis dipsaceus, L.
Chariclea umbra, Hufu.
Erastria uncula, Cl.   E. pusilla, View.
Prothymia viridaria, Cl.
Euclidia mi, Cl.   E. glyphica, L.
Catocala fraxini, L.   C. adultera, Men.   C. pacta, L.
Boletobia fuliginaria, L.
Zanclognxtha grisealis, Hb.

Herminia tentacularia, L.
Pechipagon barbalis, Cl.
Bomolocha frontis, Thnb.
Hypena rostralis, L.   H. proboscidalis, L.
Tholomiges turfosalis, Wx.
Rivula sericealis, Sc.
Brephos parthenias, L.
Geometra papilionaria, L.
Phorodesma smaragdaria, F.
Fodis putata, L.
Acidalia perochraria, F.R.   A straminata, Fr.   A. pallidata, Bkh.   A. subsericeata, Hw.   A. bisetata, Hufu. A. aversata, L.   A. aversata, ab. spoliata, Stgr.   A. immorata, L.   A. rubiginata, Hufu.   A incanata, L.   A. fumata, Stph.   A. fumata, ab. simplaria, Hb.   A. immutata, L. A. strigilaria, Hb.
Zonosoma pendularia, Cl.   Z. orbicularia, Hb.
Timandra amata, L.
Rhyparia melanaria, L.
Abraxas marginata, L. A. marginata, ab. pollutaria, Hb.
Cabera pusaria, L.   C. exanthemata, Sc.
Numeria pulveraria, L.
Ellopia prosapiaria, L.
Eugonia alniaria, L.
Selenia bilunaria, Esp.   S. lunaria, Schiff.   S. tetralunaria, Hufu.
Pericallia syringaria, L.
Odontopera bidentata. Cl.
Crocallis elinguaria, L.
Angerona prunaria, L. A. prunaria, ab. sordiata, Tuessl.
Rumina luteolata, L.
Epione apiciaria, Schiff.   E. paralellaria, Schiff.
Hypoplectis adspersaria, Hb.
Venilia macularia, L.
Macaria notata, L.   M. alternaria, Hb.   M. signaria, Hb.   M. liturata, Cl.
Ploseria pulverata, Thnb.
Biston pomonarius, B.   B. hirtarius, Cl.

Amphidasis betularius, L.
Boarmia cinetaria, Schiff. B. perepandata, L. B. roboraria, Schiff. B. glabraria, Hb. B. crepuscularia, Hb. B. punctularia, Hb.
Gnophos sordaria, Thub. G. obfuscaria, Hb.
Fidonia carbonaria, Cl.
Ematurga atomaria, L. E. atonaria, ab. unicolorio.
Bupœlus piniarius, L.
Halia loricaria, Ev. H. wanaria, L. H. brunneata, Thub.
Phasiane clathrata, L.
Lythria purpuraria, L.
Ortholitha limitata, Sc.
Odezia atrata, L. O. tibiale, Esp., var. Evers mannaria. H.S.
Anaitis praeformatr, Hb. A. paludata, Thub.
Lobophora polycommata, Hb. L. carpinata, Bkh. L. L. halterata, Hufu. L. sexalisata, Hb. L. viretata, Hb.
Cheimatobia brumata, L.
Eucosmia undulata, L.
Scotosia badiata, Hb. .
Lygris prunata L. L. testata, L. L. populata, L. L. associata, Bkh.
Cidaria fulvata, Forst. C. ocellata, L. C. bicolorata, Hufu. C. variata, Schiff. C. var. ab. obeliseata, Hb. C. juniperata, L. C. miata, L. C. taeniata, Stph. C. truncata, Hufu. C. serraria, F. C. munitata, Hb. C. aptata, Hb. C. viridaria, F. C. didymata, L. C. cambrica, Curt. C. vespertaria, Bkh. C. incursata, Hb. C. fluctuata, L. C. montanata, Bkh. C. quadrifasciaria, Cl. C. ferrugata, Cl. C. fer. ab. spadicearia, Bkh. C. unidentaria, Hw. C. suffumata, Hb. C. pomoeriaria, Ev. C. designata, Rott. C. vittata, Bkh. C. dilutata, Bkh. C. caesiata, Lang. C. cuculata, Hufu. C. rivata, Hb. C. sociata, Bkh. C. unangulata, Hw. C. albicillata, L. C. hastata, L. C. has. var. subpastata, Nolck. C. tristata, L. C. luctuata, Hb. C. alchemillata. L. C. adaequata, Bkh. C. albulata, Schiff. C. testaceata, Dow. C. decolorata, Hb. C. luteata, Schiff. C. obliter-

ata, Hufu. C. bilineata, L. C. sordidata, F. C. trifasciata, Bkh. C. silaceata, Hb. C. corylata, Thnb. C. nigrofasciaria, Goez. C. sagittata, F. C. comitata, L. C. lapidata, Hb. C. tersata, Hb.
Collix sparsata, Tr.
Eupithecia oblongata, Thnb. E. venosata, F. E. linariata, F. E. pusillata, F. E. abietaria, Goez. E. togata, Hb. E. debiliata, Hb. E. rectangulata, L. E. succenturiata, L. E. subfulvata, Hw. E. nanata, Hb. E. innotata, Hufu. E. pygmeata, Hb. E. tenuiata, Hb. E. plumbeolata, Hw. E. immundata, Z. E. satyrata, Hb. E. helveticaria, B. E. castigata, Hb. E. trisignaria, H.S. E. vulgata, Hw. E. campanulata, H.S. E. albipunctata, Hw. E. minutata, Gn. E. absinthiata, Cl. E. pimpinellata, Hb. E. conterminata, Z. E. indigata, Hb. E. exiguata, Hb. E. lanceolata, Hb. E. sobrinata, Hb.
Aglossa pinguinalis, L.
Asopia glaucinalis, L.
Scoparia centuriella Schiff. S. ambigualis, Tr. S. borealis, Tugstr. S. sudetica, Z. S. truncicolella, Stt. S. murana, Curt. S. crataegella, Hb.
Eurrhypara urticata, L.
Bolys octomaculata, F., var. trigutta, Esp. B. nyctemeralis, Hb. B. nigrata, Sc. B. cingulata, L. B. purpuralis, L. B. manualis, Hb., var. septentrionalis, Tugstr. B. lutealis, Hb. B. nebulalis, Hb. B. decrepitalis, H.S. B. fuscalis, Schiff. B. terrealis, Tr. B. sambucalis, Schiff. B. prunalis, Schiff. B. pandalis, Hb. B. ruralis, Sc.
Eurycreon sticticalis, L.
Nomophila noctuella, Schiff.
Orobena straminalis, Hb.
Pionea forficalis, L.
Diasemia litterata, Sc.
Hydrocampa stagnata, Dow. H. nymphaeta, L.
Paraponyx stratiotata, L.
Cataclysta lemnata, L.
Schoenobius forficellus, Thnb. S. mucronellus, Schiff.
Crambus hamellus, Thnb. C. pascuellus, L. C. silvellus,

LEPIDOPTERA. 273

Hb. C. ericellus, Hb. C. Heringiellus, H.S. C. alienellus, Zk. C. biarmicus, Tugstr. C. pratellus, L. C. dumetellus, Hb. C. hortuellus, Hb. C. truncatellus, Zett. C. maculalis, Zett. C. falsellus, Schiff. C. myellus, Hb. C. margaritellus, Hb. C. culmellus, L. C. contaminellus Hb. C. tristellus, F. C. perlellus, Sc. C. perlellus, var. Warringtonellus, Stt.

Dioryctria abietella, Zk.
Nephopteryx rhenella, Zk.
Pempelia betula, Goez. P. fusca, Hw. P. palumbella, F
Catastia marginea, Schiff., var. auriciliella, Hb.
Hypochalcia ahenella, Zk.
Cryptoblabes bistriga, Hw.
Myelois cirrigerella, Zk. M. advenella, Zk. M. tetricella, F.
Homoeosoma nebulea, Hb. var. maritima, Fugrt. H binaevella, Hb.
Anerastia lotella, Hb.
Ephestia elutella, Hb. E. interpunctella, Hb.
Aphomia sociella, L.
Rhacodia candana, F., var. emargana, F. R. effractana, Froel.

Teras umbrana, Hb. T. hastiana, L. T. abietana, Hb. T. maccana, Tr. T. maccana, var. basalticola, Str. T. fimbriana, Thnb. T. variegana, Schiff. T. boscana, F. T. literana, L., ab. irrorana, Hb. T. niveana, F. T. lipsiana, Schff. T. sponsana, F. T. rufana, Schiff. T. Schalleriana, L. T. comparana, Hb. T. comparana, ab. comariana, Z. T. aspersana, Hb. E. ferrugana, Tr.

Tortrix podana, Sc. T. rosana, L. T. semialbana, Gn. T. corylana, F. T. ribeana, Hb. T. heperana, Schiff. T. inopiana, Hw. T. musculana, Hb. T. politana, Hw. T. ministrana, L. T. Bergmanniana, L. T. Forsterana, F. T. viburnana, F. T. paleana, Hb. T. rusticana, Tr. T. reticulana Hb. T. grotiana, F. T. gnomana, Cl. T. Gerningana, Schiff. T. prodromana, Hb.

Sciaphila osseana, Sc. S. argentana, Cl. S. Wahlbomiana, L. S. pasivana, Hb.

T

Exapate congelatella, Cl.
Cochylis hamana, L. C. cruentana, L. C. ambiguella, Hb. C. rutilana, Hb. C. Kuhlweiniana, F. C. Hartmanniana, Cl. C. aleella, Schulze. C. badiana, Hb. C. Deutschiana, Zett. C. Smeathmanniana, F. C. Richteriana, F. C. Heydeniana, H.S. C. flammeolana, Tugst. C. phaleratana, H.S C. roseana, Hw. C. purpuratana, H.S. C. Mussehliana, Tr. C. ambiguana, Froel. C. dubitana, Hb.

Retinia pinivorana, Z. R. turionana, Hb. R. Bicoliana, Schiff. R. resinella, L.

Penthina Schreberiana, L. P. salicella, L. P. semifasciana, Hw. P. corticana, Hb. P. sororculana, Zett. P. variegana, Hb. P. ochroleucana, Hb. P. dimidiana, Sodof. P. sellana, Hb. P. gentiana, Hb. P. roseomaculana, H.S. P. lediana, L. P. pyrolana, Wk. P. postremana, Z. P. turfosana, H.S. P. arbutella, L. P. mygindana, Schiff. P. rufana, Sc. P. striana, Schiff. P. Branderiana, L. P. metallicana, Hb. P. metallicana, var. irriguana, H.S. P. stibiana, Gn. P. palustrana, Z. P. Schulziana, F. P. olivana, Tr. P. Boisduoaliana, Dup. P. arcuella, Cl. P. livulana, Sc. P. umbrosana, Frr. P. urticana, Hb. P. racunana, Dup. P. lucivagana, Z. P. cespitana, Hb. P. bipunctana, F. P. trifoliana, H.S.

Aspis Uddmanniana, L.

Aphelia lanceolana, Hb. A. furfurana, Hw.

Grapholitha obumbratana, Z. G. expallidana, Hw. G. Hohenwartiana, Tr. G. graphana, Tr. G. tedella, Cl. G. demarniana, F. G. subocellana, Dow. G. nisella, Cl. G. Penkleriana, F. G. ophthalmicana, Hb. G. Gunthera Tugstr. G. solandriana, L. G. solandriana, var. sinuani, Hb. G. bilunana, Hw. G. tetraguetiana, Hw. G. immundana, F. G. crenana, Hb. G. similana, Hb. G. incarnatana, Hb. G. tripunctana, F. G. cynosbana, F. G. Pflugiana, Hw. G. cirsiana, Z. G. Brunnichiana, Froel. G. foenella, L. G. citrana, Hb. G. aspidiscana, Hb. G. hypericana, Hb. G. albersana, Hb. G. tenebrosana, Dup. G. funebrana, Tr. G. strobiella, L. G. clanculana, Tugstr.

G. cosmophorana, Tr. G. coniferana, Rtz. G. Woeberiana, G. compositella, F. G. comucopiae, Tugstr. G. duplicana, Zett. G. perlepidana, Hw. G. leguminana, Z. G. orobana, Tr. G. aurana, F.

Carpocapsa pomonella, L.

Phthoroblastis motacillana, Z. P. populana, E. P. strangulana, Tugss.

Steganoptycha ramella, L. S. pinicolana, Z. S. signatana, Dgl. S. ustomaculana, Curt. S. vacciniana, Z. S. nemorivaga, Tugst. S. ericetana, H.S. S. fractifasciana, Hw. S. quadrana, Hb. S. pygmaeana, Hb. S. mercuriana, Hb. S. Gimmerthaliana, Z.

Phoxopteryx laetana, F. P. tineana, Hb. P. biarcuana, Stph. P. uncana, Hb. P. unguicella, L. P. siculana, Hb. P. lundana, F. P. myrtillana, Tr. P. derasana, Hb.

Rhopobota naevana, Hb.

Dichrorampha petiverella, L. D. alpinana, Fr. D. agilana, Fugst. D. plumbagana, Tr. D. acuminatana, Z. D. plumbana, Sc.

Chorentis Bjerxandrella, Thnb. C. Myllerana, F.

Simaethis diana, Hb. S. oxyacanthella, L.

Talaeporia politella, O. T. pseudobombycella, Hb.

Solenobia clathrella, F. S. cembrella, L.

Lypusa maurella, F.

Diplodoma marginepunctella, Stph.

Scardia boleti, F. S. boletella, F.

Blabophanes imella, Hb. B. rusticella, Hb. B. rus. var. spilotella, Tugst.

Tinea fulvimitrella, Sodof. T. arcella, F. T. corticella, Curt. T. fraudulentella, H.S. T. arcuatella, Stt. T. granella, L. T. cloacella, Hw. T. albipunctella, Hw. T. fuliginosella, Z. T. misella, Z. T. fuscipunctella, Hw. T. pellionella, L. T. curtella, Tugot.

Phylloporia bistrigella, Hw.

Tineola biselliella, Hum.

Lampronica luzella, Hb. L. rubiella, Bjerk.

Fucurvaria pectinea, Hw. F. flavifrontella, Hein. F. capitella, Cl. F. Ochlmanniella, Tr. F. rupella, Schiff.

Nemophora Swammerdammella, L.   N. Schwarziella, Z.
N. pilulella, Hb.   N. pilella, F.   N. metaxella, Hb.
Adela fibulella, F.   A. Degeerella, L.   A. croesella, Sc.
A. cuprella, Thnb.
Nemotois metallicus, Poda.
Ochsenheimeria bisontella, Z.
Acrolepia cariosella, Tr.
Roeslerstammia Erxlebella, F.
Hyponomento evonymellus, L.
Swammerdamia compunctella, H.S.   S. variegata, Tugst.
S. caesiella, Hb.   S. nubeculella, Tugst.   S. conspersella, Tugst.
Zelleria fasciapennella, Stt.
Argyresthia rufella, Tugst.   A. conjugella, Z.   A. sorbiella, Tr.   A. Goedartella, L.   A. Brockeella, Hb.   A. illuminatella, Z.   A. aurulentella, Stt.
Cedestis Gysseleniella, Dup.   C. farinatella, Dup.
Plutella xylostella, L.   P. annulatella, Curt.
Cerostoma radiatella, Dow.   C. asperella, L.   C. falcella, Hb.   C. dentella, F.
Semioscopis strigulana, F.   S. avellanella, Hb.   .
Epigraphia Heinkellneriana, Schiff.
Exaeretia allisella, Stt.
Depressaria flavella, Hb.   D. arenella, Schiff.   D. laterella, Schiff.   D. ciniflonella, Z.   D. Alstroemeriana, Cl.   D. applana, F.   D. ciliella, Stt.   D. angelicella, Hb.   D. hepatariella, Z.   D. depresella, Hb.   D. pimpinella, Z.   D. badiella, Hb.   D. heracliana, D.G.   D. chaerophylli, Z.
Gelechia muscosella, Z.   G. incomptella, H.S.   G. distinctella Z.   G. velocella, Dup.   G. ericetella, Hb.   G. infernalis, H.S.   G. lentiginosella, Z.   G. galbanella, Z.   G. boreella, Dgl.   G. continuella, Z.   G. longicornis, Curt.   G. diffinis, Hw.   G. lugubrella, F.   G. viduella, F.   G. luctuella, Hb.
Brachmia Mouffetella, Schiff.
Bryotropha terrella, Hb.   B. senectella, Z.   B. flavipalpella, Tugst.   B. cinerosella, Tugst.   B. obscurecinerea, Nolek.

Lita artemisiella, Tr. L. atriplicella, F. L. murinella, H.S.
L. ingloriella, Hein. L. maculea, Hw. L. maculiferella,
Dgl. L. marmorea, Hw. L. leucomelanella, Z.
Teleia proximella, Hb. T, notatella, Hb. T. triparella,
Z. T. dodecella, L.
Argyritis pictella, Z. A. superbella, Z.
Nannodia Hermannella, F.
Parasia lapella, L. P. neuropterella, Z.
Chelaria Hübnerella, Dow.
Ergatis ericinella, Dup.
Doryphora carchariella, Z. D, servella, Z. D. lutelentella, Z.
Monochroa tenebrella, Hb.
Lamprotes micella, Schiff.
Anacampsis vorticella, Sc.
Tachyptilia populella, Cl. T. temerella, Z.
Brachycrossata cinerella, Cl.
Ceratophora rufescens, Hw.
Cladodes dimidiella, Schiff., var. costiguttella, Z.
Sophronia semicostella, Hb.
Pleurota bicostella, Cl.
Hypercallia citrinalis, Sc.
Auchinia daphnella, Hb.
Dasycera sulphurella, F.
Oecophora flavifrontella, Hb. O. similella, Hb. O. cinnamomea, Z.
Hypatima binotella, Thnb.
Glyphipteryx thrasonella, Sc. G. Haworthana, Stph.
Gracilaria stigmatella, F. G, populetorum, Z. G. elongella, L. G. phasianipennella, Hb. var. auraguttella, Stph.
Ornix torguilella, Z. O. scoticella, Stt. O. scoticella, var. canella, Tugst. O. betulae, Stt.
Coleophora limosipennella, Dup. C. ocripennella, Z.
C. fuscedinella, Z. C. Binderella, Kollar. C. vimitella, Z.
C. vilisella, Gregs. C. orbitella, Z. C. gryphipennella, Bouché. C. nigricella, Stph. C. alcyonipennella, Kollar.
C. deauratella, Z. C. Fabriciella, Vill. C. anatipennella, Hb. C. therinella, Tgvtr. C. troglodytella, Dup. C.

millefolii, Z. C. punctipennella, Tugstr. C. laripenella, Zett. C. flavaginella, Z. C. murinipennella, Dup. C. caespititiella, Z.

Chauliodus Illigerellus, Hb. C. chaerophyllellus, Goez. C. idaei, Z. C. Laspeyrella, Hb. C. conturbatella, Hb. C. lacteella, Hb. C. miscella, Schiff. C. Raschkiella, Z. C. Schranckella, Hb. C. subbistrigella, Hw.

Tinagma perdicellum, Z.

Heydenia fulviguttella, Z.

Butalis obscurella, Sc. B. fallacella, Schl. B. laminella, H.S. B. chenopodiella, Hb. B. noricella, Z. B. inspersella, Hb.

Pancalia Latreillella, Curt. P. Leuwenhoekella, L.

Endrosis lacteela, Schiff.

Schreckensteinia festaliella, Hb.

Stathmopada pedella, L.

Heliozela sericiella, Hw.

Elachista apicipunctella, Stt. E. albifrontella, Hw. E. kilmunella, Stt. E helvetica, Frey. E. airea, Stt. E. obscurella Stt. E. utonella, Frey. E. rhynchosporella, Stt. E. subalbidella, Schlg.

Lithocolletis alniella, Z. L. strigulatella, Z. L. spinolella, Dup. L. pomifoliella, Z. L. sorbi, Frey. L. junoniella, Z. L. betulae, Z. L. emberizaepeunell, Bouché. L. tristrigella, Hw.

Lyonetia Clerkella, L. L. ledi, Wk. L. frigidariella, H.S.

Phylloenistis soligna, Z.

Cemiostoma susinella, H.S. C. scitella, Z.

Bucculatrix nigricomella, Z. B. crataegi Z. B. artemisiae, H.S. B. cristatella, Z.

Opostega salaciella, Tr.

Nepticula argentipedella, Z. N. Weaweri, Stt.

Micropterx calthella, L. M. aureatella, Sc. M. sparmannella, Bosc. M. chrysolepidella, Z. M. semipurpurella, Stph. M. purpurella, Hw.

Cnaenidophorus rhododactylus, F.

Platyptilia Bertrami, Roessl. P. gonodactyla, Schiff. P. Zetherstedtti, Z. P. tesseradactyla, L. P, Metzneri, Z.

Amblyptilia acanthodactyla, Hb. A. cosmadactyla, Hb.
Oxyptilus didactylus, L. O. parvidactylus, Hw.
Mimaeseopticus serotinus, F. M. pteradactylus, L.
Oedematophorus lithodactylus, Tr.
Lieoptilus scaradactylus, Hb. L. osterdactylus, Z.
Aciptilia tetradactyla, L. A. paluduni.

THE END.

www.ingramcontent.com/pod-product-compliance
Lightning Source LLC
Chambersburg PA
CBHW031340230426
43670CB00006B/397